Paul Charbonneau

Solar and Stellar Dynamos

Saas-Fee Advanced Course 39

Swiss Society for Astrophysics and Astronomy
Edited by O. Steiner

Paul Charbonneau
Département de Physique
Université de Montréal
Montréal
Canada

Volume Editor
Oskar Steiner
Kiepenheuer-Institut für Sonnenphysik
Freiburg
Germany

This Series is edited on behalf of the Swiss Society for Astrophysics and Astronomy:
Société Suisse d'Astrophysique et d'Astronomie Observatoire de Genève, ch. des Maillettes 51, CH-1290 Sauverny, Switzerland

Cover picture: Snapshot of a global, three-dimensional, numerical simulation of solar magneto-convection. The color scale codes the radial magnetic field at the outer surface of the domain: red-yellow for positive magnetic polarity and blue-cyan for negative. At the given time instant of the magnetic cycle, polarity reversal of the polar cap magnetic field is just taking place. Even though the surface magnetic field is spatiotemporally intermittent, a well-defined axisymmetric component develops with a full magnetic cycle lasting about 78 years in this specific simulation. (From the web-page of the solar physics research group at Montreal University http://www.astro.umontreal.ca/ ∼ paulchar/grps/)

ISSN 1861-7980   ISSN 1861-8227   (electronic)
ISBN 978-3-642-43047-3   ISBN 978-3-642-32093-4   (eBook)
DOI 10.1007/978-3-642-32093-4
Springer Heidelberg New York Dordrecht London

© Springer-Verlag Berlin Heidelberg 2013
Softcover reprint of the hardcover 1st edition 2013
This work is subject to copyright. All rights are reserved by the Publisher, whether the whole or part of the material is concerned, specifically the rights of translation, reprinting, reuse of illustrations, recitation, broadcasting, reproduction on microfilms or in any other physical way, and transmission or information storage and retrieval, electronic adaptation, computer software, or by similar or dissimilar methodology now known or hereafter developed. Exempted from this legal reservation are brief excerpts in connection with reviews or scholarly analysis or material supplied specifically for the purpose of being entered and executed on a computer system, for exclusive use by the purchaser of the work. Duplication of this publication or parts thereof is permitted only under the provisions of the Copyright Law of the Publisher's location, in its current version, and permission for use must always be obtained from Springer. Permissions for use may be obtained through RightsLink at the Copyright Clearance Center. Violations are liable to prosecution under the respective Copyright Law.
The use of general descriptive names, registered names, trademarks, service marks, etc. in this publication does not imply, even in the absence of a specific statement, that such names are exempt from the relevant protective laws and regulations and therefore free for general use.
While the advice and information in this book are believed to be true and accurate at the date of publication, neither the authors nor the editors nor the publisher can accept any legal responsibility for any errors or omissions that may be made. The publisher makes no warranty, express or implied, with respect to the material contained herein.

Springer is part of Springer Science+Business Media (www.springer.com)

Preface

Magnetic fields are seen to interweave the fabric of the universe wherever telescopes are directed to. They are observed and measured in such diverse astrophysical objects and scale-sizes as galaxies, interstellar clouds, stars, planets, and compact objects, including the Sun and the Earth. While primordial magnetic fields may have been pre-existent in the early universe, it is believed that most cosmic magnetic fields that we observe nowadays have been, and still are, continuously created by an electromagnetic inductive process known as the dynamo effect. Various types of astrophysical dynamos form the cosmic power plants for the production of magnetic fields. They are a central theme of virtually all kinds of astrophysical research and of planetary physics and geophysics. Here, they are the central topic this book is devoted to.

This monograph is a thorough introduction to astrophysical dynamos, leading up to present-day research questions. It is directed at physics students at the intermediate to advanced level with basic knowledge of hydro- and electrodynamics, for self-study or for study in classes. Also, the advanced astrophysicists may learn to appreciate it as an introduction to, or a reference book on, the foundations of astrophysical dynamos. After familiarizing the reader with the concept and meaning of the magnetohydrodynamic equations, it starts with basic examples of kinematic flows and discusses conditions for the onset of dynamo action. The main part of the book is devoted to the solar dynamo, which is the best studied stellar dynamo with the most precise and longest lasting observational record. Different types of stellar dynamos across the H-R diagram and a brief outlook on compact objects, galaxies, and beyond, complete the book. Each chapter is complemented with an annotated bibliography leading the reader to background information including historical matters, as well as to the more advanced literature and present-day research articles. The book is accompanied by a course Web page, which can be found under http://obswww.unige.ch/SSAA/sf39/dynamos. It contains animations of several figures of Chaps. 2 and 3, to which references are made in the text. Other than ad hoc introduced equations and definitions, all equations can be derived step by step (and should be rederived by the reader for a healthy exercise), often by using the appendices, which contain all the necessary identities and theorems from vector calculus, and operators and fluid equations in the cylindrical and spherical coordinate systems.

The present book is an extended and updated version of the lectures delivered under the same title by Paul Charbonneau during the 39th "Saas-Fee Advanced Course", which took place in Les Diablerets (Switzerland) from March 23–28, 2009. The course title was "Magnetic Fields of Stars: From the Sun to Compact Objects" and included additional lectures by Sami K. Solanki (Magnetic fields in the atmospheres of the Sun and stars) and Christopher Thompson (Stellar Magnetofluids). Sixty students from 16 countries attended the course and could benefit from the excellent lectures delivered by these top experts. The Eurotel in Les Diablerets provided much appreciated hospitality. Snow, sunshine, and not least skiing and curling were also appropriately enjoyed by the course participants.

The organizers of the course, M. Liebendörfer and F.-K. Thielemann (University of Basel), S. Berdyugina and O. Steiner (Kiepenheuer-Institut, Freiburg i.Br.), and R. Hirschi (Keele University) thank the three speakers for high-standard lectures, Paul Charbonneau for preparing the present lecture notes, and the course students for their active participation and enthusiasm. They also gratefully acknowledge financial assistance provided by the Swiss Society for Astrophysics and Astronomy.

Freiburg i. Br., May 2012   Oskar Steiner

Acknowledgments

The Lecture Notes making up this volume are an expanded and updated version of the notes distributed to attendees of the 39th Saas Fee Advanced School of the Swiss Astronomical Society. These Notes actually have had a long and tortuous genesis. They began as class notes for a graduate course in solar physics which, initially thanks to Fran Bagenal, I had the opportunity co-teach at the University of Colorado in Boulder with my then-HAO-colleague Tom Bogdan. Already with the first instance of the course, we opted to treat a few subjects at great depth, rather than give a broad, necessarily shallower survey of all of solar physics. Our emphasis was on modeling, mathematical and computational, the idea being that a detailed understanding of modeling in the context of a few topics, solidly grounded in the underlying physics, could be transferred to other subject areas and remain useful even if some of the chosen topics were to fall out fashion with time. The solar dynamo was one of these topics (which has not yet fallen out of fashion!), for which I was lead author and lecturer. We did have the privilege to have Fausto Cattaneo and David Galloway contributing lectures and written material, and the 1997 instance of the course also benefited greatly from feedback by Nic Brummell and Steve Tobias, who were attending ex officio. It is truly a pleasure for me to acknowledge here the influence all these individuals—as well as Keith MacGregor and Peter Gilman at HAO/NCAR—have had on my own thinking about dynamos.

Tom and I eventually started turning these CU class notes into a book, but as we each went our own way, geographically and professionally, this project has yet to come to fruition. Upon moving to the Université de Montréal, I used the dynamo part of this embryonic book to teach graduate courses on astrophysical magneto-hydrodynamics, adding and modifying much of the material in the process. When I was invited to lecture at Saas-Fee 39 on the topic of astrophysical dynamos, this provided a natural starting point to produce the Lecture Notes the reader now has in hand, with some additional material borrowed from an extensive review article I wrote a few years ago for the online Journal *Living Reviews in Solar Physics*, and skillful LATEX ing and editing manoeuvres by Oskar Steiner.

At various points in time, many colleagues have kindly provided numerical data or actual plots that are now included in these Lecture Notes; their contributions are explicitly acknowledged in the captions to the corresponding figures. Many graduate students, too numerous to list here, have also provided useful feedback over the years. I do wish to mention explicitly Michel-André Vallières-Nollet, Étienne Racine, Patrice Beaudoin, Alexandre St-Laurent-Lemerle and Caroline Dubé, who have gone well beyond the call of duty in verifying and/or rederiving mathematical expressions, uncovering a number of significant mistakes (usually minus signs) in the process. Naturally, I retain full responsibility for any remaining errors.

It remains for me to thank once again the organizers of the 39th Saas-Fee Advanced Course for giving me the opportunity to lecture in front of a very engaging crowd of graduate students and researchers, and, overall, for an extremely pleasant winter week in Les Diablerets.

Montréal, August 2011   Paul Charbonneau

Contents

1 Magnetohydrodynamics 1
   1.1   The Fluid Approximation 2
       1.1.1   Matter as a Continuum 2
       1.1.2   Solid Versus Fluid 3
   1.2   Essentials of Hydrodynamics........................... 4
       1.2.1   Mass: The Continuity Equation 4
       1.2.2   The D/Dt Operator 6
       1.2.3   Linear Momentum: The Navier–Stokes Equations 7
       1.2.4   Angular Momentum: The Vorticity Equation........ 10
       1.2.5   Energy: The Entropy Equation 12
   1.3   The Magnetohydrodynamical Induction Equation........... 13
   1.4   Scaling Analysis 15
   1.5   The Lorentz Force 18
   1.6   Joule Heating 20
   1.7   The Full Set of MHD Equations 20
   1.8   MHD Waves....................................... 22
   1.9   Magnetic Energy 23
   1.10 Magnetic Flux Freezing and Alfvén's Theorem 24
   1.11 The Magnetic Vector Potential 25
   1.12 Magnetic Helicity 26
   1.13 Force-Free Magnetic Fields........................... 26
   1.14 The Ultimate Origin of Astrophysical Magnetic Fields........ 27
       1.14.1 Why B and not E? 27
       1.14.2 Monopoles and Batteries....................... 28
   1.15 The Astrophysical Dynamo Problem(s).................. 30
       1.15.1 A Simple Dynamo 30
       1.15.2 The Challenges 33
   Bibliography.. 34

2 Decay and Amplification of Magnetic Fields ... 37
- 2.1 Resistive Decays of Magnetic Fields ... 37
  - 2.1.1 Axisymmetric Magnetic Fields ... 38
  - 2.1.2 Poloidal Field Decay ... 39
  - 2.1.3 Toroidal Field Decay ... 41
  - 2.1.4 Results for a Magnetic Diffusivity Varying with Depth ... 42
  - 2.1.5 Fossil Stellar Magnetic Fields ... 43
- 2.2 Magnetic Field Amplification by Stretching and Shearing ... 44
  - 2.2.1 Hydrodynamical Stretching and Field Amplification ... 44
  - 2.2.2 The Vainshtein & Zeldovich Flux Rope Dynamo ... 46
  - 2.2.3 Hydrodynamical Shearing and Field Amplification ... 48
  - 2.2.4 Toroidal Field Production by Differential Rotation ... 48
- 2.3 Magnetic Field Evolution in a Cellular Flow ... 52
  - 2.3.1 A Cellular Flow Solution ... 52
  - 2.3.2 Flux Expulsion ... 57
  - 2.3.3 Digression: The Electromagnetic Skin Depth ... 59
  - 2.3.4 Timescales for Field Amplification and Decay ... 60
  - 2.3.5 Flux Expulsion in Spherical Geometry: Axisymmetrization ... 62
- 2.4 Two Anti-Dynamo Theorems ... 64
  - 2.4.1 Zeldovich's Theorem ... 65
  - 2.4.2 Cowling's Theorem ... 66
- 2.5 The Roberts Cell Dynamo ... 68
  - 2.5.1 The Roberts Cell ... 68
  - 2.5.2 Dynamo Solutions ... 69
  - 2.5.3 Exponential Stretching and Stagnation Points ... 72
  - 2.5.4 Mechanism of Field Amplification in the Roberts Cell ... 73
  - 2.5.5 Fast Versus Slow Dynamos ... 74
- 2.6 The CP Flow and Fast Dynamo Action ... 75
  - 2.6.1 Dynamo Solutions ... 76
  - 2.6.2 Fast Dynamo Action and Chaotic Trajectories ... 78
  - 2.6.3 Magnetic Flux Versus Magnetic Energy ... 80
  - 2.6.4 Fast Dynamo Action in the Nonlinear Regime ... 81
- 2.7 Dynamo Action in Turbulent Flows ... 82
- Bibliography ... 83

3 Dynamo Models of the Solar Cycle ... 87
- 3.1 The Solar Magnetic Field ... 88
  - 3.1.1 Sunspots and the Photospheric Magnetic Field ... 88
  - 3.1.2 Hale's Polarity Laws ... 90
  - 3.1.3 The Magnetic Cycle ... 92

		3.1.4	Sunspots as Tracers of the Sun's Internal Magnetic Field	93
		3.1.5	A Solar Dynamo Shopping List	94
	3.2	Mean-Field Dynamo Models		95
		3.2.1	Mean-Field Electrodynamics	95
		3.2.2	The α-Effect	97
		3.2.3	Turbulent Pumping	101
		3.2.4	The Turbulent Diffusivity	102
		3.2.5	The Mean-Field Dynamo Equations	103
		3.2.6	Dynamo Waves	103
		3.2.7	The Axisymmetric Mean-Field Dynamo Equations	105
		3.2.8	Linear $\alpha\Omega$ Dynamo Solutions	107
		3.2.9	Nonlinearities and α-Quenching	112
		3.2.10	Kinematic $\alpha\Omega$ Models with α-Quenching	113
		3.2.11	Enters Meridional Circulation: Flux Transport Dynamos	116
		3.2.12	Interface Dynamos	118
	3.3	Babcock–Leighton Models		121
		3.3.1	Sunspot Decay and the Babcock–Leighton Mechanism	122
		3.3.2	Axisymmetrization Revisited	127
		3.3.3	Dynamo Models Based on the Babcock–Leighton Mechanism	128
		3.3.4	The Babcock–Leighton Poloidal Source Term	129
		3.3.5	A Sample Solution	130
	3.4	Models Based on HD and MHD Instabilities		132
		3.4.1	Models Based on Shear Instabilities	132
		3.4.2	Models Based on Flux-Tube Instabilities	134
	3.5	Global MHD Simulations		135
	3.6	Local MHD Simulations		143
	Bibliography			146
4	**Fluctuations, Intermittency and Predictivity**			**153**
	4.1	Observed Patterns of Solar Cycle Variations		153
		4.1.1	Pre-Telescopic and Early Telescopic Sunspot Observations	153
		4.1.2	The Sunspot Cycle	155
		4.1.3	The Butterfly Diagram	156
		4.1.4	The Waldmeier and Gnevyshev–Ohl Rules	158
		4.1.5	The Magnetic Activity Cycle	160
		4.1.6	The Maunder Minimum	160
		4.1.7	From Large-Scale Magnetic Fields to Sunspot Number	162

	4.2	Cycle Modulation Through Stochastic Forcing	164
	4.3	Cycle Modulation Through the Lorentz Force	168
	4.4	Cycle Modulation Through Time Delays	171
	4.5	Intermittency	173
	4.6	Model-based Cycle Predictions	176
		4.6.1 The Solar Polar Magnetic Field as a Precursor	177
		4.6.2 Model-Based Prediction Using Solar Data	180
	Bibliography		182
5	**Stellar Dynamos**		187
	5.1	Early-Type Stars	189
		5.1.1 Mean-Field Models	189
		5.1.2 Numerical Simulations of Core Dynamo Action	192
		5.1.3 Getting the Magnetic Field to the Surface	194
		5.1.4 Alternative to Core Dynamo Action	194
	5.2	A-Type Stars	195
		5.2.1 Observational Overview	195
		5.2.2 The Fossil Field Hypothesis	197
		5.2.3 Dynamical Stability of Large-Scale Magnetic Fields	197
		5.2.4 The Transition to Solar-Like Dynamo Activity	197
	5.3	Solar-Type Stars	198
		5.3.1 Observational Overview	198
		5.3.2 Empirical Stellar Activity Relationships	199
		5.3.3 Solar and Stellar Spin-Down	200
		5.3.4 Modelling Dynamo Action in Solar-Type Stars	206
	5.4	Fully Convective Stars	207
	5.5	Pre- and Post-Main-Sequence Stars	208
	5.6	Compact Objects	209
	5.7	Galaxies and Beyond	210
	Bibliography		211

Appendix A: Useful Identities and Theorems from Vector Calculus ... 215

Appendix B: Coordinate Systems and the Fluid Equations ... 217

Appendix C: Physical and Astronomical Constants ... 227

Appendix D: Maxwell's Equations and Physical Units ... 229

Index ... 233

List of Previous Saas-Fee Advanced Courses

!! 2006 First Light in the Universe
    A. Loeb, A. Ferrara, R. S. Ellis
!! 2005 Trans-Neptunian Objects and Comets
    D. Jewitt, A. Morbidelli, H. Rauer
!! 2004 The Sun, Solar Analogs and the Climate
    J. D. Haigh, M. Lockwood, M. S. Giampapa
!! 2003 Gravitational Lensing: Strong, Weak and Micro
    P. Schneider, C. Kochanek, J. Wambsganss
!! 2002 The Cold Universe
    A. Blain, F. Combes B. Draine
!! 2001 Extrasolar Planets
    P. Cassen, T. Guillot, A. Quirrenbach
!! 2000 High-Energy Spectroscopic Astrophysics
    S. M. Kahn, P. von Ballmoos, R. A. Sunyaev
!! 1999 Physics of Star Formation in Galaxies
    F. Palla H. Zinnecker
!! 1998 Star Clusters
    B. W. Carney, W. E. Harris
!! 1997 Computational Methods for Astrophysical Fluid Flow
    R. J. LeVeque, D. Mihalas, E. A. Dorfi, E. Müller
!! 1996 Galaxies: Interactions and Induced Star Formation
    R. C. Kennicutt, Jr., F. Schweizer, J. E. Barnes
!! 1995 Stellar Remnants
    S. D. Kawaler, I. Novikov, G. Srinivasan
!! 1994 Plasma Astrophysics
    J. G. Kirk, D. B. Melrose, E. R. Priest

!! 1993 The Deep Universe
   A. R. Sandage, R. G. Kron, M. S. Longair
!! 1992 Interacting Binaries
   S. N. Shore, M. Livio, E. P. J. van den Heuvel
!! 1991 The Galactic Interstellar Medium
   W. B. Burton, B. G. Elmegreen, R. Genzel
!! 1990 Active Galactic Nuclei
   R. D. Blandford, H. Netzer, L. Woltjer
* 1989 The Milky Way as a Galaxy
   G. Gilmore, I. King, P. van der Kruit
! 1988 Radiation in Moving Gaseous Media
   H. Frisch, R. P. Kudritzki, H. W. Yorke
! 1987 Large Scale Structures in the Universe
   A. C. Fabian, M. Geller, A. Szalay
! 1986 Nucleosynthesis and Chemical Evolution
   J. Audouze, C. Chiosi, S. E. Woosley
! 1985 High Resolution in Astronomy
   R. S. Booth, J. W. Brault, A. Labeyrie
! 1984 Planets, Their Origin, Interior and Atmosphere
   D. Gautier, W. B. Hubbard, H. Reeves
! 1983 Astrophysical Processes in Upper Main Sequence Stars
   A. N. Cox, S. Vauclair, J. P. Zahn
* 1982 Morphology and Dynamics of Galaxies
   J. Binney, J. Kormendy, S. D. M. White
! 1981 Activity and Outer Atmospheres of the Sun and Stars
   F. Praderie, D. S. Spicer, G. L. Withbroe
* 1980 Star Formation
   J. Appenzeller, J. Lequeux, J. Silk
* 1979 Extragalactic High Energy Physics
   F. Pacini, C. Ryter, P. A. Strittmatter
* 1978 Observational Cosmology
   J. E. Gunn, M. S. Longair, M. J. Rees
* 1977 Advanced Stages in Stellar Evolution
   I. Iben Jr., A. Renzini, D. N. Schramm
* 1976 Galaxies
   K. Freeman, R. C. Larson, B. Tinsley
* 1975 Atomic and Molecular Processes in Astrophysics
   A. Dalgarno, F. Masnou-Seeuws, R. V. P. McWhirter
* 1974 Magnetohydrodynamics
   L. Mestel, N. O. Weiss
* 1973 Dynamical Structure and Evolution of Stellar Systems
   G. Contopoulos, M. Hénon, D. Lynden-Bell

List of Previous Saas-Fee Advanced Courses

* 1972 Interstellar Matter
    N. C. Wickramasinghe, F. D. Kahn, P. G. Metzger
* 1971 Theory of the Stellar Atmospheres
    D. Mihalas, B. Pagel, P. Souffrin

!! May be ordered from Springer and/or are available online at springerlink.com

! May be ordered from
  Geneva Observatory
  Saas-Fee Courses
  CH-1290 Sauverny, Switzerland

* Out of print

** In preparation

Chapter 1
Magnetohydrodynamics

> *From a long view of history—seen from, say, ten thousand years from now—there can be little doubt that the most significant event of the 19th century will be judged as Maxwell's discovery of the laws of electrodynamics.*
>
> Richard Feynman
> The Feynman Lectures on Physics, vol. II (1964)

To sum it all up in a single sentence, *magnetohydrodynamics* (hereafter MHD) is concerned with the behavior of electrically conducting but globally neutral fluids flowing at non-relativistic speeds and obeying Ohm's law. Remarkably, most astrophysical fluids meet these apparently stringent requirements, the most glaring exception being the relativistic inflows and outflows powered by compact objects such as black holes or neutron stars.

The focus of these lectures is on the amplification of solar and stellar magnetic fields through the inductive action of fluid flows, a process believed to be well-described by MHD for physical conditions characterizing the interior of the sun and (most) stars. Before we dive into MHD proper, we will first clarify what we mean by "fluid" (Sect. 1.1), and review the fundamental physical laws governing the flow of unmagnetized fluid, i.e., classical hydrodynamics (Sect. 1.2). We then introduce magnetic fields into the fluid picture (Sects. 1.3–1.13), and close by reflecting upon the ultimate origin of astrophysical magnetic fields (Sect. 1.14), and establishing the various incarnations of the so-called dynamo problem (Sect. 1.15) which will occupy our attention in the subsequent chapters.

1.1 The Fluid Approximation

1.1.1 Matter as a Continuum

It did take some two thousand years to figure it out, but we now know that Democritus was right after all: matter is composed of small, microscopic "atomic" constituents. Yet on our daily macroscopic scale, things sure look smooth and continuous. Under what circumstances can an assemblage of microscopic elements be treated as a continuum? The key constraint is that there be a good *separation of scales* between the "microscopic" and "macroscopic".

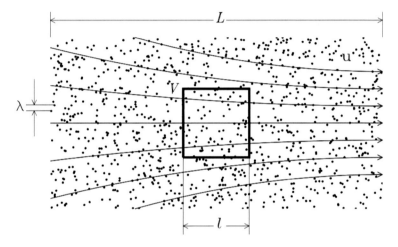

Fig. 1.1 Microscopic view of a fluid. In general the velocity of microscopic constituents is comprised of two parts: a randomly-oriented thermal velocity, and a systematic drift velocity, which, on the macroscopic scale amounts to what we call a flow u. A fluid representation is possible if the mean inter-particle distance λ is much smaller than the global length scale L.

Consider the situation depicted on Fig. 1.1, corresponding to an amorphous substance (spatially random distribution of microscopic constituents). Denote by λ the mean interparticle distance, and by L the macroscopic scale of the system; we now seek to construct macroscopic variables defining fluid characteristics at the macroscopic scale. For example, if we are dealing with an assemblage of particles of mass m, then the *density* (ϱ) associated with a cartesian volume element of linear dimensions l centered at position x would be given by something like:

$$\varrho(x) = \frac{1}{l^3} \sum_k m_k \quad [\text{kg m}^{-3}], \tag{1.1}$$

where the sum runs over all particles contained within the volume element. One often hears or reads that for a continuum representation to hold, it is only necessary that

1.1 The Fluid Approximation

Table 1.1 Spatial scales of some astrophysical objects and flows

System/flow	ϱ [kg/m^3]	N [m^{-3}]	λ [m]	L [km]
Solar interior	100	10^{29}	10^{-10}	10^5
Solar atmosphere	10^{-4}	10^{23}	10^{-8}	10^3
Solar corona	10^{-11}	10^{17}	10^{-6}	10^5
Solar wind (1 AU)	10^{-21}	10^7	0.006	10^5
Molecular cloud	10^{-20}	10^7	0.001	10^{14}
Interstellar medium	10^{-21}	10^6	0.01	10^{16}

the particle density be "large". But how large, and large with respect to what? For the above expression to yield a well-defined quantity, in the sense that the numerical value of ϱ so computed does not depend sensitively on the size and location of the volume element, or on time if the particles are moving, it is essential that a great many particles be contained within the element. Moreover, if we want to be writing differential equations describing the evolution of ϱ, the volume element better be infinitesimal, in the sense that it is much smaller that the macroscopic length scale over which global variables such as ϱ may vary. These two requirements translate in the double inequality:

$$\lambda \ll l \ll L \ . \tag{1.2}$$

Because the astrophysical systems and flows that will be the focus of our attention span a very wide range of macroscopic sizes, the continuum/fluid representation will turn out to hold in circumstances where the density is in fact minuscule, as you can verify for yourself upon perusing the collection of astrophysical systems listed in Table 1.1 below.[1] In all cases, a very good separation of scales does exist between the microscopic (λ) and macroscopic (L).

1.1.2 Solid Versus Fluid

Most continuous media can be divided into two broad categories, namely *solids* and *fluids*. The latter does not just include the usual "liquids" of the vernacular, but also gases and plasmas. Physically, the distinction is made on the basis of a medium's response to an applied *stress*, as illustrated on Fig. 1.2. A volume element of some continuous substance is subjected to a shear stress, i.e., two force acting tangentially and in opposite directions on two of its parallel bounding surface (black arrows). A

---

[1] All density-related estimate assume a gas of fully ionized Hydrogen ($\mu = 0.5$) for the sun, of neutral Hydrogen for the interstellar medium ($\mu = 1$), and molecular Hydrogen ($\mu = 2$) for molecular clouds. Solar densities are for the base of the convection zone (solar interior), optical depth unity (atmosphere), and typical coronal loop (corona). N is the number density of microscopic constituents. The length scale listed for the solar atmosphere is the granulation dimension, for the corona it is the length of a coronal loop, for the solar wind the size of Earth's magnetosphere, and that for the interstellar medium is the thickness of the galactic (stellar) disk; all rounded to the nearest factor of ten.

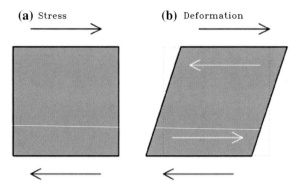

Fig. 1.2 Deformation of a mass element in response to a stress pattern producing a horizontal shear (*black arrows*). A solid will rapidly reach an equilibrium where internal stresses (*white arrows*) produced by the deformation will equilibrate the applied shear. A fluid at rest cannot generate internal stresses, and so will be increasingly deformed for as long as the external shear is applied.

solid will immediately generate a restoring force (white arrows), ultimately due to electrostatic interactions between its microscopic constituents, and vigorously resist deformation (try shearing a brick held between the palms of your hands!). The solid will rapidly reach a new equilibrium state characterized by a finite deformation, and will relax equally rapidly to its initial state once the external stress vanishes. A *fluid*, on the other hand, can offer no resistance to the applied stress, at least in the initial stages of the deformation.

1.2 Essentials of Hydrodynamics

The governing principles of classical hydrodynamics are the same as those of classical mechanics, transposed to continuous media: conservation of mass, linear momentum, angular momentum and energy. The fact that these principles must now be applied not to point-particles, but to spatially extended and deformable volume elements (which may well be infinitesimal, but they are still finite!) introduces some significant complications, mostly with regards to the manner in which forces act. Let's start with the easiest of our conservation statements, that for mass, as it exemplifies very well the manner in which conservation laws are formulated in moving fluids.

1.2.1 Mass: The Continuity Equation

Consider the situation depicted on Fig. 1.3, namely that of an arbitrarily shaped fictitious surface S fixed in space and enclosing a volume V embedded in a fluid of density $\varrho(x)$ moving with velocity $u(x)$. The *mass flux* associated with the flow across the (closed) surface is

1.2 Essentials of Hydrodynamics

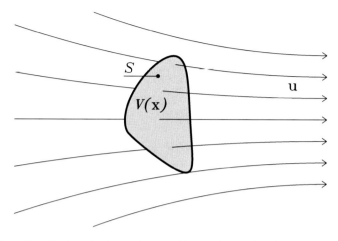

Fig. 1.3 An arbitrarily shaped volume element V bounded by a closed surface S, both fixed in space, and traversed by a flow $\boldsymbol{u}$.

$$\Phi = \oint_S \varrho \boldsymbol{u} \cdot \hat{\boldsymbol{n}} \, dS \quad [\mathrm{kg\, s^{-1}}], \tag{1.3}$$

where $\hat{\boldsymbol{n}}$ is a unit vector everywhere perpendicular to the surface, and by convention oriented towards the exterior. The mass of fluid contained within V is simply

$$M = \int_V \varrho \, dV \quad [\mathrm{kg}]. \tag{1.4}$$

This quantity will evidently vary if the mass flux given by Eq. (1.3) is non-zero:

$$\frac{\partial M}{\partial t} = -\Phi. \tag{1.5}$$

Here the minus sign is a direct consequence of the exterior orientation of $\hat{\boldsymbol{n}}$. Inserting Eqs. (1.3) and (1.4) into (1.5) and applying the divergence theorem to the RHS of the resulting expression yields:

$$\frac{\partial}{\partial t} \int_V \varrho \, dV = -\int_V \nabla \cdot (\varrho \boldsymbol{u}) \, dV. \tag{1.6}$$

Because V is fixed in space, the $\partial/\partial t$ and $\int_V$ operators commute, so that

$$\int_V \left[\frac{\partial \varrho}{\partial t} + \nabla \cdot (\varrho \boldsymbol{u}) \right] dV = 0. \tag{1.7}$$

Because V is completely arbitrary, in general this can only be satisfied provided that

$$\boxed{\frac{\partial \varrho}{\partial t} + \nabla \cdot (\varrho \boldsymbol{u}) = 0}. \tag{1.8}$$

This expresses mass conservation in differential form, and is known in hydrodynamics as the *continuity equation*.

Incompressible fluids have constant densities, so that in this limiting case the continuity equation reduces to

$$\nabla \cdot \boldsymbol{u} = 0 \quad \text{[incompressible]}. \tag{1.9}$$

Water is perhaps the most common example of an effectively incompressible fluid (under the vast majority of naturally occurring conditions anyway). The gaseous nature of most astrophysical fluids may lead you to think that incompressibility is likely to be a pretty lousy approximation in cases of interest in this course. It turns out that the incompressibility can lead to a pretty good approximation of the behavior of compressible fluids provided that the flow's Mach number (ratio of flow speed to sound speed) is much smaller than unity.

The density of a fluid can vary by means other than mechanical forcing; in many astrophysical circumstances, thermal dilation effects are in fact the primary driver of density variations. Most fluids will dilate when heated, a property measured by the *coefficient of thermal dilation*:

$$\alpha = -\frac{1}{\varrho}\frac{\partial \varrho}{\partial T} \quad [\text{K}^{-1}]. \tag{1.10}$$

Water, for example, has a substantial coefficient of thermal dilation of $\simeq 10^{-4}\,\text{K}^{-1}$ at $T = 10\,°\text{C}$. Thermal expansion of sea water, rather than meltdown of the polar ice caps, is in fact the primary contributor to sea level rise associated with predicted climate change for the coming century.

In an environment stratified by gravity, such as the atmospheres and interiors of the sun and stars, a localized heat input can then, via thermal dilation, lead to the appearance of a buoyancy force through a decrease in density of the heated volume element. If not equilibrated sufficiently quickly by thermal diffusion, this will generate a flow known as *thermal convection*, which will play an important role in all solar/stellar dynamo models considered in later chapters.

1.2.2 The D/Dt Operator

Suppose we want to compute the time variation of some physical quantity (Z, say) at some fixed location $\boldsymbol{x}_0$ in a flow $\boldsymbol{u}(\boldsymbol{x})$. In doing so we must take into account the fact that Z is in general both an explicit and implicit function of time, because the volume element "containing" Z is moving with the fluid, i.e., $Z \to Z(t, \boldsymbol{x}(t))$. We therefore need to use the chain rule and write:

1.2 Essentials of Hydrodynamics

$$\frac{dZ}{dt} = \frac{\partial Z}{\partial t} + \frac{\partial Z}{\partial x}\frac{\partial x}{\partial t} + \frac{\partial Z}{\partial y}\frac{\partial y}{\partial t} + \frac{\partial Z}{\partial z}\frac{\partial z}{\partial t}. \tag{1.11}$$

Noting that $\boldsymbol{u} = d\boldsymbol{x}/dt$, this becomes

$$\frac{dZ}{dt} = \frac{\partial Z}{\partial t} + \frac{\partial Z}{\partial x}u_x + \frac{\partial Z}{\partial y}u_y + \frac{\partial Z}{\partial z}u_z = \frac{\partial Z}{\partial t} + (\boldsymbol{u} \cdot \nabla)Z. \tag{1.12}$$

This corresponds to the time variation of Z *following the fluid element as it is carried by the flow*. It is a very special kind of derivative in hydrodynamics, known as the *Lagrangian derivative*, which will be represented by the operator:

$$\boxed{\frac{D}{Dt} \equiv \frac{\partial}{\partial t} + (\boldsymbol{u} \cdot \nabla)}. \tag{1.13}$$

Note in particular that the Lagrangian derivative of $\boldsymbol{u}$ yields the acceleration of a fluid element:

$$\boldsymbol{a} = \frac{D\boldsymbol{u}}{Dt}, \tag{1.14}$$

a notion that will soon come very handy when we write $F = ma$ for a fluid.

A *material surface* is defined as an ensemble of points that define a surface, all moving along with the flow. Therefore, in a local frame of reference S' co-moving with any infinitesimal element of a material surface, $\boldsymbol{u}' = 0$. The distinction between material surfaces, as opposed to surfaces fixed in space such as in Eq. (1.3), has crucial consequences with respect to the commuting properties of temporal and spatial differential operators. In the latter case $\int_V$ commutes with $\partial/\partial t$, whereas for material surfaces and volume elements it is D/Dt that commutes with $\int_V$ (and $\oint_S$, etc.).

1.2.3 Linear Momentum: The Navier–Stokes Equations

A force $\boldsymbol{F}$ acting on a point-object of mass m is easy to deal with; it simply produces an acceleration $\boldsymbol{a} = \boldsymbol{F}/m$ in the same direction as the force (sounds simple but it still took the genius of Newton to figure it out...). In the presence of a force acting on the surface of a spatially extended fluid element, the resulting fluid acceleration will depend on both the orientation of the force and the surface. We therefore define the net force $\boldsymbol{t}$ in terms of a *stress tensor*:

$$t_x = \hat{\boldsymbol{e}}_x s_{xx} + \hat{\boldsymbol{e}}_y s_{xy} + \hat{\boldsymbol{e}}_z s_{xz}, \tag{1.15}$$
$$t_y = \hat{\boldsymbol{e}}_x s_{yx} + \hat{\boldsymbol{e}}_y s_{yy} + \hat{\boldsymbol{e}}_z s_{yz}, \tag{1.16}$$
$$t_z = \hat{\boldsymbol{e}}_x s_{zx} + \hat{\boldsymbol{e}}_y s_{zy} + \hat{\boldsymbol{e}}_z s_{zz}, \tag{1.17}$$

where "s_{xy}" denotes the force per unit area acting in the x-direction on a surface perpendicular to the y-direction, t_x is the net force acting in the x-direction, and similarly for the other components. Consider now a unit vector perpendicular to a surface arbitrarily oriented in space:

$$\hat{n} = \hat{e}_x n_x + \hat{e}_y n_y + \hat{e}_z n_z, \quad n_x^2 + n_y^2 + n_z^2 = 1. \quad (1.18)$$

The net force along this direction is simply

$$t_{\hat{n}} = (\hat{n} \cdot \hat{e}_x) t_x + (\hat{n} \cdot \hat{e}_y) t_y + (\hat{n} \cdot \hat{e}_z) t_z = \hat{n} \cdot \mathsf{s}. \quad (1.19)$$

We can now use the Lagrangian acceleration to write "$a = F/m$" for a fluid element occupying a volume V bounded by a surface S:

$$\frac{D}{Dt} \int_V u \, dV = \frac{1}{\varrho} \oint_S \mathsf{s} \cdot \hat{n} \, dS, \quad (1.20)$$

where the LHS represents the mean acceleration of the fluid element, and ϱ on the RHS is its mean density. We now pull the same tricks as in Sect. 1.2.1: use the divergence theorem to turn the surface integral into a volume integral, commute the temporal derivative and volume integral on the LHS, invoke the arbitrariness of the actual integration volume V, and finally make good use of the continuity equation (1.8), to obtain the differential equation for u:

$$\frac{Du}{Dt} = \frac{1}{\varrho} \nabla \cdot \mathsf{s} \quad [\mathrm{m\,s^{-2}}]. \quad (1.21)$$

We now define the *pressure* (units: pascal; $1\,\mathrm{Pa} \equiv 1\,\mathrm{N\,m^{-2}}$) as the isotropic part of the force acting perpendicularly on the volume's surfaces, and separate it explicitly from the stress tensor:

$$\mathsf{s} = -p\mathbf{I} + \boldsymbol{\tau}, \quad (1.22)$$

where $\mathbf{I}$ is the identity tensor, and the minus sign arises from the convention that pressures acts on the bounding surface towards the interior of the volume element, and $\boldsymbol{\tau}$ will presently become the *viscous stress tensor*. Since $\nabla \cdot (p\mathbf{I}) = \nabla p$, Eq. (1.21) becomes

$$\boxed{\frac{Du}{Dt} = -\frac{1}{\varrho} \nabla p + \frac{1}{\varrho} \nabla \cdot \boldsymbol{\tau}}. \quad (1.23)$$

This is the celebrated *Navier–Stokes equation*. Any additional volumetric body forces (gravity, Lorentz force, etc.) are simply added to the RHS.

The next step is to obtain expressions for the components of the tensor $\boldsymbol{\tau}$. The viscous force, which is what $\boldsymbol{\tau}$ stands for, can be viewed as a form of friction acting between contiguous laminae of fluid moving with different velocities, so that we

1.2 Essentials of Hydrodynamics

expect it to be proportional to velocity *derivatives*. Consider now the following decomposition of a velocity gradient:

$$\frac{\partial u_k}{\partial x_l} = \underbrace{\frac{1}{2}\left(\frac{\partial u_k}{\partial x_l} + \frac{\partial u_l}{\partial x_k}\right)}_{D_{kl}} + \underbrace{\frac{1}{2}\left(\frac{\partial u_k}{\partial x_l} - \frac{\partial u_l}{\partial x_k}\right)}_{\Omega_{kl}}. \quad (1.24)$$

The first term on the RHS is a *pure shear*, and is described by the (symmetric) *deformation tensor* D_{kl}; the second is a *pure rotation*, and is described by the antisymmetric *vorticity tensor* Ω_{kl}. It can be shown that the latter causes no deformation of the fluid element, *therefore the viscous force can only involve* D_{kl}. A *Newtonian fluid* is one for which the (tensorial) relation between τ and D_{kl} is linear:

$$\tau_{ij} = f_{ij}(D_{kl}), \quad i, j, k, l = (1, 2, 3) \equiv (x, y, z). \quad (1.25)$$

Since τ and D are both symmetric tensors, this linear relationship can involve up to 36 independent numerical coefficients. The next step is to invoke the invariance of the physical laws embodied in Eq. (1.25) under rotation of the coordinate axes to set some of these coefficients to zero. The mathematics is rather tedious, but worth the effort because at the end of the day you end up with:

$$\tau_{xx} = 2\mu D_{xx} + (\zeta - \frac{2}{3}\mu)(D_{xx} + D_{yy} + D_{zz}), \quad (1.26)$$

$$\tau_{yy} = 2\mu D_{yy} + (\zeta - \frac{2}{3}\mu)(D_{xx} + D_{yy} + D_{zz}), \quad (1.27)$$

$$\tau_{zz} = 2\mu D_{zz} + (\zeta - \frac{2}{3}\mu)(D_{xx} + D_{yy} + D_{zz}), \quad (1.28)$$

$$\tau_{xy} = 2\mu D_{xy}, \quad (1.29)$$

$$\tau_{yz} = 2\mu D_{yz}, \quad (1.30)$$

$$\tau_{zx} = 2\mu D_{zx}, \quad (1.31)$$

which now involves only two numerical coefficients, μ and ζ, known as the *dynamical viscosity* and *bulk viscosity*, respectively. It is often convenient to define a coefficient of *kinematic viscosity* as

$$\nu = \frac{\mu}{\varrho} \quad [\text{m}^2\,\text{s}^{-1}]. \quad (1.32)$$

In an incompressible flow, the terms multiplying ζ vanish and it is possible to rewrite the Navier–Stokes equation in the simpler form:

$$\frac{D\boldsymbol{u}}{Dt} = -\frac{1}{\varrho}\nabla p + \nu\nabla^2\boldsymbol{u} \quad [\text{incompressible}]. \quad (1.33)$$

Note here the presence of a Laplacian operator acting on a *vector* quantity (here $\boldsymbol{u}$); this is only equivalent to the Laplacian acting on the scalar components of $\boldsymbol{u}$ in the special case of cartesian coordinates.[2]

Incompressible or not, the behavior of viscous flows will often hinge on the relative importance of the advective and dissipative terms in the Navier–Stokes equation:

$$\varrho(\boldsymbol{u} \cdot \nabla)\boldsymbol{u} \quad \leftrightarrow \quad \nabla \cdot \boldsymbol{\tau} . \tag{1.34}$$

Introducing characteristic length scales u_0, L, ϱ_0 and ν_0, dimensional analysis yields:

$$\varrho_0 \frac{u_0^2}{L} \quad \leftrightarrow \quad \frac{1}{L} \varrho_0 \nu_0 \frac{u_0}{L} , \tag{1.35}$$

where we made use of the fact that the viscous stress tensor has dimensions $\mu \times D_{ik}$, with $\mu = \varrho\nu$ and the deformation tensor D_{ik} has dimension of velocity per unit length (cf. Eq. 1.24). The ratio of these two terms is a dimensionless quantity called the *Reynolds Number*:

$$\boxed{\mathrm{Re} = \frac{u_0 L}{\nu_0}} . \tag{1.36}$$

This measures the importance of viscous forces versus fluid inertia. It is a key dimensionless parameter in hydrodynamics, as it effectively controls fundamental processes such as the transition to turbulence, as well as more mundane matters such as boundary layer thicknesses.

A few words on boundary conditions; in the presence of viscosity, the flow speed must vanish wherever the fluid is in contact with a rigid surface S:

$$\boldsymbol{u}(\boldsymbol{x}) = 0 , \quad \boldsymbol{x} \in S . \tag{1.37}$$

This remains true even in the limit where the viscosity is vanishingly small. For a *free surface* (e.g., the surface of a fluid sphere floating in a vacuum), the normal components of both the flow speed and viscous stress must vanish instead:

$$\boldsymbol{u} \cdot \hat{\boldsymbol{n}}(\boldsymbol{x}) = 0 , \quad \boldsymbol{\tau} \cdot \hat{\boldsymbol{n}} = 0 , \quad \boldsymbol{x} \in S . \tag{1.38}$$

1.2.4 Angular Momentum: The Vorticity Equation

The "rotation" and "angular momentum" of a fluid system cannot be reduced to simple scalars such as angular velocity and moment of inertia, because the application

---

[2] See Appendix B for full listing of differential operators in cylindrical and spherical polar coordinates.

1.2 Essentials of Hydrodynamics

of a torque to a fluid element can alter not just its rotation rate, but also its shape and mass distribution. A more useful measure of "rotation" is the *circulation* Γ about some closed contour γ embedded in and moving with the fluid:

$$\Gamma(t) = \oint_\gamma \boldsymbol{u}(\boldsymbol{x},t) \cdot d\boldsymbol{\ell} = \int_S (\nabla \times \boldsymbol{u}) \cdot \hat{\boldsymbol{n}} \, dS = \int_S \boldsymbol{\omega} \cdot \hat{\boldsymbol{n}} \, dS \,, \tag{1.39}$$

where the second equality follows from Stokes' theorem, and the third from the definition of *vorticity*:

$$\boldsymbol{\omega} = \nabla \times \boldsymbol{u} \,. \tag{1.40}$$

Thinking about flows in terms of vorticity $\boldsymbol{\omega}$ rather than speed $\boldsymbol{u}$ can be useful because of *Kelvin's theorem*, which states that in the inviscid limit $\nu \to 0$ (or, equivalently, $\mathrm{Re} \to \infty$), the circulation Γ along any closed loop γ advected by the moving fluid is a conserved quantity:

$$\frac{D\Gamma}{Dt} = 0 \,. \tag{1.41}$$

Applying Eq. (1.39) yields the equivalent expression

$$\boxed{\frac{D}{Dt} \int_S \boldsymbol{\omega} \cdot \hat{\boldsymbol{n}} \, dS = 0} \,, \tag{1.42}$$

stating that the flux of vorticity across any material surface S bounded by γ is also a conserved quantity, both in fact being integral expressions of angular momentum conservation.

An evolution equation for $\boldsymbol{\omega}$ can be obtained via the Navier–Stokes equation, in a particularly illuminating manner in the case of an incompressible fluid ($\nabla \cdot \boldsymbol{u} = 0$) with constant kinematic viscosity ν, in which case Eq. (1.33) can be rewritten as

$$\frac{D\boldsymbol{u}}{Dt} = -\nabla\left(\frac{p}{\varrho} + \Phi\right) - \nu \nabla \times (\nabla \times \boldsymbol{u}) \quad \text{[incompressible]}\,, \tag{1.43}$$

where it was assumed that gravity can be expressed as the gradient of a (gravitational) potential. Taking the curl on each side of this expression then yields:

$$\nabla \times \left(\frac{\partial \boldsymbol{u}}{\partial t}\right) + \nabla \times (\boldsymbol{u} \cdot \nabla \boldsymbol{u}) = \underbrace{-\nabla \times \left[\nabla\left(\frac{p}{\varrho} + \Phi\right)\right]}_{=0} - \nu \nabla \times \nabla \times (\nabla \times \boldsymbol{u}) \,, \tag{1.44}$$

then, commuting the time derivative with $\nabla\times$ and making judicious use of some vector identities to develop the second term on the LHS, remembering also that $\nabla \cdot \boldsymbol{\omega} = 0$, eventually leads to:

$$\boxed{\frac{D\omega}{Dt} - \omega \cdot \nabla u = \nu \nabla^2 \omega \quad \text{[incompressible]}} \quad . \tag{1.45}$$

This is the *vorticity equation*, expressing in differential form the conservation of the fluid's angular momentum. A useful vorticity-related quantity is the *kinetic helicity*, defined as

$$h = u \cdot \omega , \tag{1.46}$$

which measures the amount of twisting in a flow. This will prove an important concept when investigating magnetic field amplification by fluid flows.

1.2.5 Energy: The Entropy Equation

Omitting, to begin with, the energy dissipated in heat by viscous friction, the usual accounting of energy flow into and out of a volume element V fixed in space leads to the following differential equation expressing the conservation of the plasma's *internal energy* per unit mass (e, in units J/kg):

$$\frac{De}{Dt} + (\gamma - 1)e \nabla \cdot u = \frac{1}{\varrho} \nabla \cdot \left[(\chi + \chi_r) \nabla T \right], \tag{1.47}$$

where for a perfect gas we have

$$e = \frac{1}{\gamma - 1} \frac{p}{\varrho} = \frac{1}{\gamma - 1} \frac{kT}{\mu m} , \tag{1.48}$$

with $\gamma = c_p/c_v$ the ratio of specific heats, and $(\chi + \chi_r)\nabla T$ the heat flux in or out of the fluid element, with χ and χ_r the coefficients of thermal and radiative conductivity, respectively (units: $\mathrm{J\,K^{-1}\,m^{-1}\,s^{-1}}$). Equation (1.47) expresses that any variation of the specific energy in a plasma volume moving with the flow (LHS) is due to heat flowing in or out of the volume by conduction or radiation (here in the diffusion approximation). The "extra" term $\propto \nabla \cdot u$ on the LHS of Eq. (1.47) embodies the work done against (or by) the pressure force in compressing (or letting expand) the volume element.

It is often convenient to rewrite the energy conservation equation in terms of the plasma's *entropy* $S \propto \varrho^{-\gamma} p$, which allows to express Eq. (1.47) in the more compact form:

$$\boxed{\varrho T \frac{DS}{Dt} = \nabla \cdot \left[(\chi + \chi_r) \nabla T \right]} , \tag{1.49}$$

which states, now unambiguously, that any change in the entropy S as one follows a fluid element (LHS) can only be due to heat flowing out of or into the domain by conduction or radiation (RHS).

1.2 Essentials of Hydrodynamics

While this is seldom an important factor in astrophysical flows, in general we must add to the RHS of Eq. (1.49) the heat produced by viscous dissipation (and, as we shall see later, by Ohmic dissipation). This is given by the so-called (volumetric) *viscous dissipation function*:

$$\phi_\nu = \frac{\mu}{2}\left(\frac{\partial u_i}{\partial x_k} + \frac{\partial u_k}{\partial x_i} - \frac{2}{3}\delta_{ik}\frac{\partial u_s}{\partial x_s}\right)^2 + \zeta\left(\frac{\partial u_s}{\partial x_s}\right)^2 \quad [\text{J}\,\text{m}^{-3}\text{s}^{-1}], \quad (1.50)$$

where summation over repeated indices is implied here. Note that since ϕ_ν is positive definite, its presence on the RHS of Eq. (1.49) can only increase the fluid element's entropy, which makes perfect sense since friction, which is what viscosity is for fluids, is an irreversible process.

For more on classical hydrodynamics, see the references listed in the bibliography at the end of this chapter.

1.3 The Magnetohydrodynamical Induction Equation

Our task is now to generalize the governing equations of hydrodynamics to include the effects of the electric and magnetic fields, and to obtain evolution equations for these two physical quantities. Keep in mind that electrical charge neutrality, as required by MHD, does not imply that the fluid's microscopic constituents are themselves neutral, but rather that positive and negative electrical charges are present in equal numbers in any fluid element.

The starting point, you guess it I hope, is Maxwell's celebrated equations:

$$\nabla \cdot \boldsymbol{E} = \frac{\varrho_e}{\varepsilon_0} \quad [\text{Gauss' law}], \quad (1.51)$$

$$\nabla \cdot \boldsymbol{B} = 0 \quad [\text{Anonymous}], \quad (1.52)$$

$$\nabla \times \boldsymbol{E} = -\frac{\partial \boldsymbol{B}}{\partial t} \quad [\text{Faraday's law}], \quad (1.53)$$

$$\nabla \times \boldsymbol{B} = \mu_0 \boldsymbol{J} + \mu_0\varepsilon_0 \frac{\partial \boldsymbol{E}}{\partial t} \quad [\text{Ampère's/Maxwell's law}], \quad (1.54)$$

where, in the SI system of units, the electric field is measured in $\text{N}\,\text{C}^{-1}$ ($\equiv \text{V}\,\text{m}^{-1}$), and the magnetic field[3] $\boldsymbol{B}$ in tesla (T). The quantity ϱ_e is the electrical charge density (C m^{-3}), and $\boldsymbol{J}$ is the electrical current density (A m^{-2}). The permittivity ε_0 (= 8.85×10^{-12} C^2 N^{-1}m^{-2} in vacuum) and magnetic permeability μ_0 (= $4\pi \times 10^{-7}$ N A^{-2} in vacuum) can be considered as constants in what follows, since we will not be dealing with polarisable or ferromagnetic substances.

---

[3] Strictly speaking, $\boldsymbol{B}$ should be called the magnetic flux density or somesuch, but on this one we'll stick to common astrophysical usage.

The first step is (with all due respect to the man) to do away altogether with Maxwell's displacement current in Eq. (1.54). This can be justified if the fluid flow is non-relativistic and there are no batteries around being turned on or off, two rather sweeping statements that will be substantiated in Sect. 1.5. For the time being we just revert to the original form of Ampère's law:

$$\nabla \times \boldsymbol{B} = \mu_0 \boldsymbol{J} \, . \tag{1.55}$$

In general, the application of an electric field $\boldsymbol{E}$ across an electrically conducting substance will generate an electrical current density $\boldsymbol{J}$. Ohm's law postulates that the relationship between $\boldsymbol{J}$ and $\boldsymbol{E}$ is linear:

$$\boldsymbol{J}' = \sigma \boldsymbol{E}' \, , \tag{1.56}$$

where σ is the electrical conductivity (units: $C^2 s^{-1} m^{-3} kg^{-1} \equiv \Omega^{-1} m^{-1}$, $\Omega \equiv$ Ohm). Here the primes ("''") are added to emphasize that Ohm's law is expected to hold in a conducting substance *at rest*. In the context of a fluid that moves with velocity $\boldsymbol{u}$ (relativistic or not), Eq. (1.56) can only be expected to hold in a reference frame comoving with the fluid. So we need to transform Eq. (1.56) to the laboratory (rest) frame. In the non-relativistic limit ($u/c \ll 1$, implying $\gamma \to 1$), the usual Lorentz transformation for the electrical current density simplifies to $\boldsymbol{J}' = \boldsymbol{J}$, and that for the electric field to $\boldsymbol{E}' = \boldsymbol{E} + \boldsymbol{u} \times \boldsymbol{B}$, so that Ohm's law takes on the generalized form

$$\boxed{\boldsymbol{J} = \sigma(\boldsymbol{E} + \boldsymbol{u} \times \boldsymbol{B})} \, , \tag{1.57}$$

or, making use of the pre-Maxwellian form of Ampère's law and reorganizing the terms:

$$\boldsymbol{E} = -\boldsymbol{u} \times \boldsymbol{B} + \frac{1}{\mu_0 \sigma}(\nabla \times \boldsymbol{B}) \, . \tag{1.58}$$

We now insert this expression for the electric field into Faraday's law (1.53) to obtain the justly famous *magnetohydrodynamical induction equation*:

$$\boxed{\frac{\partial \boldsymbol{B}}{\partial t} = \nabla \times (\boldsymbol{u} \times \boldsymbol{B} - \eta \nabla \times \boldsymbol{B})} \, , \tag{1.59}$$

where

$$\eta = \frac{1}{\mu_0 \sigma} \quad [m^2 s^{-1}] \tag{1.60}$$

is the *magnetic diffusivity*.[4] The first term on the RHS of Eq. (1.59) represents the inductive action of fluid flowing across a magnetic field, while the second term represents dissipation of the electrical currents sustaining the field.

Keep in mind that any solution of Eq. (1.59) must also satisfy Eq. (1.52) at all times. It can be easily shown (try it!) that if $\nabla \cdot \boldsymbol{B} = 0$ at some initial time, the form of Eq. (1.59) guarantees that zero divergence will be maintained at all subsequent times.[5]

1.4 Scaling Analysis

The evolution of a magnetic field under the action of a prescribed flow $\boldsymbol{u}$ will depend greatly on whether or not the inductive term on the RHS of Eq. (1.59) dominates the diffusive term. Under what conditions will this be the case? We seek a first (tentative) answer to this question by performing a dimensional analysis of Eq. (1.59); this involves replacing the temporal derivative by $1/\tau$ and the spatial derivatives by $1/\ell$, where τ and ℓ are time and length scales that suitably characterizes the variations of both $\boldsymbol{u}$ and $\boldsymbol{B}$:

$$\frac{B}{\tau} = \frac{u_0 B}{\ell} + \frac{\eta B}{\ell^2}, \tag{1.61}$$

where B and u_0 are a "typical" values for the flow velocity and magnetic field strength over the domain of interest. The ratio of the first to second term on the RHS of Eq. (1.61) is a dimensionless quantity known as the *magnetic Reynolds number*[6]:

$$\boxed{R_m = \frac{u_0 \ell}{\eta}}, \tag{1.62}$$

which measures the relative importance of induction versus dissipation *over length scales of order* ℓ. Note that R_m does not depend on the magnetic field strength, a direct consequence of the linearity (in $\boldsymbol{B}$) of the MHD induction equation. Our scaling analysis simply says that in the limit $R_m \gg 1$, induction by the flow dominates the evolution of $\boldsymbol{B}$, while in the opposite limit of $R_m \ll 1$, induction makes a negligible contribution and $\boldsymbol{B}$ simply decays away under the influence of Ohmic dissipation.

One may anticipate great simplifications of magnetohydrodynamics if we operate in either of these limits. If $R_m \ll 1$, only the second term is retained on the RHS of Eq. (1.61), which leads immediately to

---

[4] A note of warning: some MHD textbooks use the symbol "η" for the inverse conductivity (units Ω m), so that the dissipative term on the RHS of the induction equation retains a μ_0^{-1} prefactor.

[5] This is true under exact arithmetic; if numerical solutions to Eq. (1.59) are sought, care must be taken to ensure $\nabla \cdot \boldsymbol{B} = 0$ as the solution is advanced in time.

[6] Note the structural similarity with the usual viscous Reynolds number defined in Sect. 1.2.3, with the magnetic diffusivity η replacing the kinematic viscosity ν in the denominator. Had we not absorbed μ_0 in our definition of η, the magnetic permeability μ_0 would appear in the numerator of the magnetic Reynolds number, which I personally find objectionable.

$$\boxed{\tau = \frac{\ell^2}{\eta}}, \tag{1.63}$$

a quantity known as the *magnetic diffusion time*. It measures the time taken for a magnetic field contained in a volume of typical linear dimension ℓ to dissipate and/or diffusively leak out of the volume. Now, for most astrophysical objects, this timescale turns out to be quite large (see Table 1.2), indeed often larger than the age of the universe! This is not so much because astrophysical plasmas are such incredibly good electrical conductors—copper at room temperature is much better in this respect— but rather because astrophysical objects tend to be very, very large. The *existence* of solar and stellar magnetic fields is then not really surprising; any large-scale fossil field present in a star's interior upon its arrival on the ZAMS would still be there today at almost its initial strength. The challenge in modelling the solar and stellar magnetic fields is to reproduce the peculiarities of their spatiotemporal variations, most notably the decadal cyclic variations observed in the sun and solar-type stars.

The opposite limit $R_m \gg 1$, defines the *ideal MHD* limit. Then it is the first term that is retained on the RHS of Eq. (1.61), so that

$$\tau = \ell/u_0, \tag{1.64}$$

corresponding to the *turnover time* associated with the flow $\boldsymbol{u}$. Note already that under ideal MHD, the only non-trivial (i.e., $\boldsymbol{u} \neq 0$ and $\boldsymbol{B} \neq 0$) steady-state ($\partial/\partial t = 0$) solutions of the MHD equation with $\boldsymbol{B}$ vanishing at infinity are only possible for field-aligned flows.

Table 1.2 below lists estimates of the magnetic Reynolds number (and related physical quantities) for the various astrophysical systems considered earlier in Table 1.1.[7] The magnetic Reynolds number is clearly huge in all cases, which would suggest that the ideal MHD limit is the one most applicable to all these astrophysical systems. But things are not so simple. From a purely mathematical point of view, taking the limit $R_m \to \infty$ of the MHD induction equation is problematic, because the order of the highest spatial derivatives decreases by one. This situation is similar to the behavior of viscous flows at very high Reynolds number: solutions to Eq. (1.59) with $\eta \to 0$ in general *do not* smoothly tend towards solutions obtained for $\eta = 0$. Moreover, the distinction between the two physical regimes $R_m \ll 1$ and $R_m \gg 1$ is meaningful as long as one can define a suitable R_m for the flow as a whole, which, in turn, requires one to estimate, a priori, a length scale ℓ that adequately characterizes the flow and magnetic field at all time and throughout the spatial domain of interest. As we proceed it will become clear that this is not always straightforward, or

---

[7] Choices for length scale ℓ ($\equiv L$) as in Table 1.1. Velocity estimates correspond to large convective cells (solar interior), granulation (photosphere), solar wind speed (corona and solar wind), and turbulence (molecular clouds and interstellar medium). All these numbers (especially the turbulent velocity estimates) are again very rough, and rounded to the nearest factor of ten. The magnetic diffusivity estimates given for molecular clouds and interstellar medium depend critically on the assumed degree of ionization, and so are also very rough.

1.4 Scaling Analysis

Table 1.2 Properties of some astrophysical objects and flows

System/flow	L [km]	σ [Ω^{-1} m^{-1}]	η [m^2 s^{-1}]	τ [yr]	u [km/s^1]	R_m
Solar interior	10^6	10^4	100	10^9	0.1	10^9
Solar atmosphere	10^3	10^3	1000	10^2	1	10^6
Solar corona	10^5	10^6	1	10^8	10	10^{12}
Solar wind (1 AU)	10^5	10^4	100	10^8	300	10^{11}
Molecular cloud	10^{14}	10^2	10^4	10^{17}	100	10^{18}
Interstellar medium	10^{16}	10^3	1000	10^{22}	100	10^{21}
Sphere of copper	10^{-3}	10^8	10^{-1}	10^{-7}	–	–

even possible. Finally, the scaling analysis does away entirely with the geometrical aspects of the problem, by substituting $u_0 B$ for $\boldsymbol{u} \times \boldsymbol{B}$; yet there are situations (e.g., a field-aligned flow) where even a very large $\boldsymbol{u}$ has no inductive effect whatsoever.

Another important dimensionless quantity to be encountered in subsequent chapters is the *magnetic Prandtl number*, equal to the ratio of the viscous Reynolds number to the magnetic Reynolds number, or, equivalently, of viscosity over magnetic diffusivity:

$$P_m = \frac{\nu}{\eta}. \qquad (1.65)$$

For physical conditions characteristic of non-degenerate stellar interiors, the use of microscopic values for ν and η yields P_m in the range 0.1–0.01.

In magnetohydrodynamics, the magnetic field is supported by an electrical current density, as embodied in Ampère's law (Eq. 1.55). Because the plasma is assumed electrically neutral, this current density must arise from a drift speed $\boldsymbol{v}$ between oppositely charged microscopic constituents. The associated electrical current density is then

$$\boldsymbol{J} = nq\boldsymbol{v}, \qquad (1.66)$$

where n and q are the number density and charge of the drifting particles.

Now, dimensional analysis of Ampère's law gives an estimate of the current density required to sustain a magnetic field of strength B varying over a length scale L: $|\boldsymbol{J}| \sim B/\mu_0 L \simeq 10^{-5}$ A m^{-2} for a solar-like mean surface magnetic field of strength 10^{-3} T pervading a sphere of solar radius 7×10^8 m. Substituting this value in Eq. (1.66) and assuming a fully ionized proton+electron mixture of mean density 10^2 kg m^{-3} (appropriate for the base of the solar convection zone, at depth $r/R \simeq 0.7$), one can compute a mean drift speed that turns out to be absolutely minuscule, i.e., $|\boldsymbol{v}| \sim 10^{-15}$ m s^{-1}. This is a direct reflection of the very large number of charge carriers available to sustain the electrical current density, e.g., $n \sim 10^{29}$ m^{-3} at depth $r/R = 0.7$ in the interior of the sun. This is also why the single-fluid MHD approximation works so well in this context. Induction, as represented by the $\boldsymbol{u} \times \boldsymbol{B}$ term in the induction Eq. (1.59), results from a variation of this drift speed caused by the action of the Lorentz force on individual charged constituents mechanically forced to flow across a pre-existing magnetic field.

1.5 The Lorentz Force

Getting to Eq. (1.59) was pretty easy because we summarily swept the displacement current under the rug, but it represents only half (in fact the easy half) of our task; we must now investigate the effect of the magnetic field on the flow u; and this, it turns out, is the tricky part of the MHD approximation.

You will certainly recall that the *Lorentz force* acting on a particle carrying an electrical charge q and moving at velocity u in a region of space permeated by electric and magnetic fields is given by

$$f = q(E + u \times B) \quad [\text{N}], \tag{1.67}$$

where q is the electrical charge. Consider now a volume element ΔV containing many such particles; in the continuum limit, the total force per unit volume (F) acting on the volume element will be the sum of the forces acting on each individual charged constituent divided by the volume element:

$$\begin{aligned} F &= \frac{1}{\Delta V} \sum_k f_k = \frac{1}{\Delta V} \sum_k q_k (E + u_k \times B) \\ &= \left(\frac{1}{\Delta V} \sum_k q_k \right) E + \left(\frac{1}{\Delta V} \sum_k q_k u_k \right) \times B \\ &= \varrho_e E + J \times B \quad [\text{N m}^{-3}], \end{aligned} \tag{1.68}$$

where the last equality follows from the usual definition of charge density and electrical current density. At this point you might be tempted to eliminate the term proportional to E, on the grounds that in MHD we are dealing with a globally neutral plasma, meaning $\varrho_e = 0$, therefore $\varrho_e E \equiv 0$ and that's the end of it. That would be way too easy...

Let's begin by taking the divergence on both side of the generalized form of Ohm's law (Eq. 1.57). We then make use of Gauss's law (Eq. 1.51) to get rid of the $\nabla \cdot E$ term, and of the charge conservation law

$$\frac{\partial \varrho_e}{\partial t} + \nabla \cdot J = 0 \tag{1.69}$$

to get rid of the $\nabla \cdot J$ term. The end result of all this physico-algebraical juggling is the following expression:

$$\frac{\partial \varrho_e}{\partial t} + \frac{\varrho_e}{(\varepsilon_0/\sigma)} + \sigma \nabla \cdot (u \times B) = 0. \tag{1.70}$$

The combination ε_0/σ has units of time, and is called the *charge relaxation time*, henceforth denoted τ_e. It is the timescale on which charge separation takes place in a

1.5 The Lorentz Force

conductor if an electric field is suddenly turned on. For most conductors, this a very small number, of order 10^{-18} s !! This is because the electrical field reacts to the motion of electric charges at the speed of light (in the substance under consideration, which is slower than in a vacuum but still mighty fast). Indeed, in a conducting fluid at rest ($u = 0$) the above expression integrates readily to

$$\varrho_e(t) = \varrho_e(0)\exp(-t/\tau_e), \qquad (1.71)$$

thus the name "relaxation time" for τ_e.

Now let us consider the case of a slowly moving fluid, in the sense that it is moving on a timescale much larger than τ_e; this means that the induced electrical field will vary on a similar timescale (at best), and therefore the time derivative of ϱ_e can be neglected in comparison to the ϱ_e/τ_e term in Eq. (1.70), leading to

$$\varrho_e = \varepsilon_0 \nabla \cdot (u \times B). \qquad (1.72)$$

This indicates that a finite charge density can be sustained inside a *moving* conducting fluid. The associated electrostatic force per unit volume, $\varrho_e E$, is definitely non-zero but turns out to much smaller than the magnetic force. Indeed, a dimensional analysis of Eq. (1.68), using Eq. (1.72) to estimate ϱ_e, gives:

$$\varrho_e E \sim \left(\frac{\varepsilon_0 u B}{\ell}\right)\left(\frac{J}{\sigma}\right) \sim \left(\frac{u\tau_e}{\ell}\right) JB, \qquad (1.73)$$

$$J \times B \sim JB, \qquad (1.74)$$

where Ohm's law was used to express E in terms of J, and once again ℓ is a typical length scale characterizing the variations of the flow and magnetic field. The ratio of electrostatic to magnetic forces is thus of order $u\tau_e/\ell$. Now $\tau_e \ll 1$ to start with, and for non-relativistic fluid motion we can expect that the flow's turnover time ℓ/u is much larger than the crossing time for an electromagnetic disturbance $\sim \ell/c \sim \tau_e$; both effects conspire to render the electrostatic force absolutely minuscule compared to the magnetic force, so that Eq. (1.68) becomes

$$\boxed{F = J \times B \quad \text{[MHD approximation]}}. \qquad (1.75)$$

This must be added to the RHS of the Navier–Stokes equation (1.23)... with a $1/\varrho$ prefactor so we get a force per unit mass, rather than per unit volume.

Now, getting back to this business of having dropped the displacement current in the full Maxwellian form of Ampère's law (Eq. 1.54); it can now be all justified on the grounds that the time derivative of the charge density can be neglected in the non-relativistic limit. Indeed, to be consistent the charge conservation equation (1.69) now reduces to

$$\nabla \cdot \boldsymbol{J} = 0 \, ; \tag{1.76}$$

taking the divergence on both sides of Eq. (1.54) then leads to

$$\nabla \cdot \boldsymbol{J} = -\varepsilon_0 \nabla \cdot \left(\frac{\partial \boldsymbol{E}}{\partial t} \right) = \varepsilon_0 \frac{\partial}{\partial t} (\nabla \cdot \boldsymbol{E}) = \frac{\partial \varrho_e}{\partial t} \, . \tag{1.77}$$

This demonstrates that dropping the time derivative of the charge density is equivalent to neglecting Maxwell's displacement current in Eq. (1.54). To sum up, provided we exclude very rapid transient events (such as turning a battery on or off, or any such process which would generate a large $\partial \varrho_e / \partial t$), under the MHD approximation the following statements are all equivalent:

- The fluid motions are non-relativistic;
- The electrostatic force can be neglected as compared to the magnetic force;
- Maxwell's displacement current can be neglected.

1.6 Joule Heating

In the presence of finite electrical conductivity, the volumetric heating associated with the dissipation of electric currents must be included on the RHS of the energy equation, in the form of the so-called *Joule heating function*:

$$\phi_B = \frac{\eta}{\mu_0} (\nabla \times \boldsymbol{B})^2 \quad [\mathrm{J\,m^{-3}s^{-1}}] \, . \tag{1.78}$$

Note however that in very nearly all astrophysical circumstances, Joule heating makes an insignificant contribution to the energy budget. When it occurs, heating by magnetic energy dissipation, such as in flares, involves dynamical mechanisms that lead to effective dissipation far more rapid and efficient than Joule heating.

1.7 The Full Set of MHD Equations

For the record, we now collect the set of partial differential equations governing the behavior of magnetized fluids in the MHD limit. In anticipation of developments to follow, we write these equations in a frame of reference rotating with angular velocity $\boldsymbol{\Omega}$, with the centrifugal force absorbed within the pressure gradient term:

1.7 The Full Set of MHD Equations

$$\frac{\partial \varrho}{\partial t} + \nabla \cdot (\varrho \boldsymbol{u}) = 0 , \tag{1.79}$$

$$\frac{D\boldsymbol{u}}{Dt} = -\frac{1}{\varrho}\nabla p - 2\boldsymbol{\Omega} \times \boldsymbol{u} + \boldsymbol{g} + \frac{1}{\mu_0 \varrho}(\nabla \times \boldsymbol{B}) \times \boldsymbol{B} + \frac{1}{\varrho}\nabla \cdot \boldsymbol{\tau} , \tag{1.80}$$

$$\frac{De}{Dt} + (\gamma - 1)e\nabla \cdot \boldsymbol{u} = \frac{1}{\varrho}\left[\nabla \cdot \left((\chi + \chi_r)\nabla T\right) + \phi_\nu + \phi_B\right] , \tag{1.81}$$

$$\frac{\partial \boldsymbol{B}}{\partial t} = \nabla \times (\boldsymbol{u} \times \boldsymbol{B} - \eta \nabla \times \boldsymbol{B}) . \tag{1.82}$$

Equations (1.79)–(1.82) are further complemented by the two constraint equations:

$$\nabla \cdot \boldsymbol{B} = 0 , \tag{1.83}$$

$$p = f(\varrho, T, \ldots) , \tag{1.84}$$

and suitable expressions for the viscous stress tensor and for the physical coefficient ν, χ, η, etc. Note that gravity $\boldsymbol{g}$ is explicitly included on the RHS of (1.80), that e is the specific internal energy of the plasma (magnetic energy will be dealt with separately shortly), and that Eq. (1.84) is just some generic form for an equation of state linking the pressure to the properties of the plasma such as density, temperature, chemical composition, etc.

This is it in principle, but in what follows we shall seldom solve these equations in this complete form. In the parameter regime characterizing most astrophysical fluids, we usually have Re $\gg$ 1, which means that the $(\boldsymbol{u} \cdot \nabla)\boldsymbol{u}$ term hidden in the Lagrangian derivative on the LHS of Eq. (1.80) will play an important role; this, in turn, means turbulence, already in itself an unsolved problem even for unmagnetized fluids. There is also a strong nonlinear coupling between Eqs. (1.80) and (1.82), so that the turbulent cascade involves both the flow and magnetic field. Finally, with both Re $\gg$ 1 and $R_m \gg$ 1, astrophysical flows will in general develop structures on length scales very much smaller than that characterizing the system under study, so that even fully numerical solutions of the above set of MHD equations will tax the power of the largest extant massively parallel computers, and will continue to do so in the foreseeable future; which is why judicious geometrical and/or physical simplification remains a key issue in the art of astrophysical magnetohydrodynamics... and will also continue to remain so in the same foreseeable future!

Nonetheless, there are a few brave souls out there who have tackled the study of thermal convection and dynamo action in the sun and stars by solving the above MHD equations, going almost as far back as the first electronic computers. We will encounter a few such simulations in later chapters. In modelling dynamo action over solar cycle timescales, the MHD equations (1.79)–(1.83) have been solved numerically under either one of two physical approximations. Under the *Boussinesq approximation* the fluid density ϱ is considered constant, except where it multiplies gravity on the RHS of the momentum equation, thus retaining the effect of thermal buoyancy. In this case the mass conservation equation (1.79) reverts to the form appropriate for an incompressible fluid, i.e., $\nabla \cdot \boldsymbol{u} = 0$.

The *anelastic approximation*, nowadays in common usage, retains the possibility that ϱ be a function of position and is thus better applicable to stratified environment, but still precludes any temporal variation of density other than associated with thermal dilation. The mass conservation equation is then replaced by $\nabla \cdot (\varrho \boldsymbol{u}) = 0$. In both cases the primary practical advantage is to filter out sound waves and the fast magnetosonic wave modes, which in turn allow the use of a much larger time step than in fully compressible MHD. Both approximations still capture the Alfvén wave mode (more on this one immediately below), and the anelastic approximation also allows the propagation of gravity waves in stably stratified environments.

1.8 MHD Waves

Although it looks innocuous enough, the magnetic force in the MHD approximation has some rather complex consequences for fluid flows, as we will have ample occasions to verify throughout this course. One particularly intricate aspects relates to the types of *waves* that can be supported in a magnetized fluid; in a classical unmagnetized fluid, one deals primarily with sound waves (pressure acting as a restoring force), gravity waves (gravity acting as restoring force), or Rossby waves (Coriolis as a restoring force). It turns out that the Lorentz force introduces not one, but really two additional restoring forces.

Making judicious use of Eqs. (1.52) and (1.55), together with some classical vector identities, Eq. (1.75) can be rewritten as

$$\boldsymbol{F} = \frac{1}{\mu_0}\left[(\boldsymbol{B}\cdot\nabla)\boldsymbol{B} - \frac{1}{2}\nabla(\boldsymbol{B}^2)\right], \tag{1.85}$$

where $\boldsymbol{B}^2 \equiv \boldsymbol{B}\cdot\boldsymbol{B}$. The first term on the RHS is the *magnetic tension*, and the second the *magnetic pressure*. Fluctuations in magnetic pressure can propagate as a longitudinal wave, much as a sound wave, as depicted on Fig. 1.4a. In fact, two such *magnetosonic* waves modes actually exist, according to whether the magnetic pressure fluctuation is in phase with the gas pressure fluctuation (the so-called fast mode), or in antiphase (the slow mode). In addition, magnetic tension can produce a restoring force that allows the propagation of wave-on-a-string-like transverse waves, known as *Alfvén waves*, as illustrated on Fig. 1.4b.

Small-amplitude Alfvén waves travel with a speed a given by

$$a = \frac{B_0}{\sqrt{\mu_0 \varrho}}, \tag{1.86}$$

where B_0 is the magnitude of the (uniform) magnetic field along which the wave is propagating, and ϱ is the (constant) fluid density. Mechanical forcing of a magnetic field permeating a compressible fluid will in general excite all three wave modes. This is an essential aspect of the development and propagation of perturbations in

1.8 MHD Waves

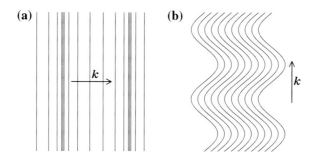

Fig. 1.4 The two fundamental MHD wave modes in a uniform background magnetic field: **a** magnetosonic mode, and **b** Alfvén mode. The wave vector k is indicated as a *thick arrow*, and highlights the fact that the magnetosonic mode is a longitudinal wave, while the Alfvén mode is a transverse wave. In the presence of plasma, the magnetosonic mode breaks into two submodes, according to the phasing between the magnetic pressure and gas pressure perturbations (see text).

the solar corona and wind, but plays a minor role in solar interior dynamics... unless one happens to be interested in subsurface magnetohelioseismology.

1.9 Magnetic Energy

Consider the expression resulting from dotting B into the induction equation (1.59), integrating over the spatial domain (V) under consideration, and making judicious use of various well-known vector identities and of Gauss' theorem:

$$\frac{d}{dt}\int_V \frac{B^2}{2\mu_0}dV = -\oint_S (S \cdot \hat{n})\,dS - \int_V (u \cdot F)\,dV - \int_V \sigma^{-1}J^2\,dV, \quad (1.87)$$

where $\hat{n}$ is a outward-directed unit vector normal to the boundary surface, and the vector quantity S is the Poynting flux:

$$S = \frac{1}{\mu_0}E \times B. \quad (1.88)$$

Examine now the three terms on the RHS of Eq. (1.87); the first is the Poynting flux component into the domain, integrated over the domain boundaries, i.e., the flux of electromagnetic energy in (integrand < 0) or out (integrand > 0) of the domain. This term evidently vanishes in the absence of applied magnetic or electric fields on the boundaries. The second is the work done by the Lorentz force (F) on the flow. In general this term can be either positive or negative; in the dynamo context we are interested in the $u \cdot F < 0$ situation, where the flow transfers energy to the magnetic field. The third term is evidently always negative, and represents the rate of energy loss due to Ohmic dissipation. Equation (1.87) then naturally leads to interpret the quantity $B^2/2\mu_0$ as the magnetic energy density, and the total *magnetic energy* ($\mathcal{E}_B$) within the domain is:

$$\mathcal{E}_B = \frac{1}{2\mu_0} \int_V B^2 \mathrm{d}V .\qquad(1.89)$$

The MHD dynamos that will be the focus of much of what follows are fluid systems that convert mechanical energy into magnetic energy, through the agency of the $u \cdot F$ term on the RHS of Eq. (1.87).

1.10 Magnetic Flux Freezing and Alfvén's Theorem

Let us return to the differential form of Faraday's law:

$$\nabla \times E = -\frac{\partial B}{\partial t} .\qquad(1.90)$$

Project now each side of this expression onto a unit vector normal to some surface S fixed in space and bounded by a closed countour γ, integrate over S, and apply Stokes' theorem to the LHS:

$$\int_S (\nabla \times E) \cdot \hat{n} \, \mathrm{d}S = \oint_\gamma E \cdot \mathrm{d}\ell = -\int_S \left(\frac{\partial B}{\partial t}\right) \cdot \hat{n} \, \mathrm{d}S .\qquad(1.91)$$

So far the surface S remains completely arbitrary. If it is fixed in space, then we get the usual integral form of Faraday's law:

$$\oint_\gamma E \cdot \mathrm{d}\ell = -\frac{\partial}{\partial t} \int_S B \cdot \hat{n} \, \mathrm{d}S ,\qquad(1.92)$$

with the LHS corresponding to the electromotive force, and the RHS to the time variation of the magnetic flux (Φ_B). If we now assume instead that the surface S is a material surface moving with the fluid, then (1) we must substitute the Lagrangian derivative D/Dt for the partial derivative on the RHS of Eq. (1.92); and (2) we are allowed to invoke Ohm's law to substitute J/σ for E on the LHS since any point of the (material) contour is by definition co-moving with the fluid:

$$\frac{1}{\sigma} \oint_\gamma J \cdot \mathrm{d}\ell = -\frac{\mathrm{D}}{\mathrm{D}t} \int_S B \cdot \hat{n} \, \mathrm{d}S .\qquad(1.93)$$

Now, obviously, in the limit of infinite conductivity we have

$$\frac{\mathrm{D}}{\mathrm{D}t} \int_S B \cdot \hat{n} \, \mathrm{d}S = 0 .\qquad(1.94)$$

1.10 Magnetic Flux Freezing and Alfvén's Theorem

This states that in the ideal MHD limit $\sigma \to \infty$, the magnetic flux threading any (open) surface is a conserved quantity as the surface is advected (and possibly deformed) by the flow. This results is known as *Alfvén's theorem*. Note in particular that in the limit of an infinitesimal surface pierced by "only one" fieldline, Alfvén's theorem is equivalent to saying that magnetic fieldlines must move in the same way as fluid elements; it is customary to say that the magnetic flux is *frozen* into the fluid. In this manner it behaves just like vorticity in the inviscid limit $\nu \to 0$. And like in the case of vorticity, sheared flows can amplify magnetic fields by stretching, a subject we will investigate in detail in the following chapter.

Alfvén's theorem can be arrived at in a different way, upon noting that the magnetic field is a solenoidal vector, in that $\nabla \cdot \boldsymbol{B} = 0$; any such vector transported by a flow $\boldsymbol{u}$ is subjected to the so-called kinematic theorem, stating that:

$$\frac{\mathrm{D}}{\mathrm{D}t} \int_{S_\mathrm{m}} \boldsymbol{B} \cdot \hat{\boldsymbol{n}} \,\mathrm{d}S = \int_{S_\mathrm{m}} \left[\frac{\partial \boldsymbol{B}}{\partial t} - \nabla \times (\boldsymbol{u} \times \boldsymbol{B}) \right] \cdot \hat{\boldsymbol{n}} \,\mathrm{d}S \,. \tag{1.95}$$

Now in the ideal limit, the RHS is zero as per our MHD induction equation (1.59) with $\eta = 0$, and the LHS is just the magnetic flux threading the material surface S_m, so there we have it.

From yet another point of view, one can also consider that what the flow $\boldsymbol{u}$ is really transporting are the current systems sustaining the magnetic field (cf. Sect. 1.4), as per Eq. (1.55). In the absence of Ohmic dissipation, these currents are moving along with the fluid without attenuation, and therefore so does the associated magnetic field.

1.11 The Magnetic Vector Potential

It will often prove useful to work with the MHD induction equation written in terms of a *vector potential*, $\boldsymbol{A}$ (units T m), such that $\boldsymbol{B} = \nabla \times \boldsymbol{A}$. Equation (1.59) is then readily recast in the form:

$$\nabla \times \left[\frac{\partial \boldsymbol{A}}{\partial t} - \boldsymbol{u} \times (\nabla \times \boldsymbol{A}) + \eta \nabla \times (\nabla \times \boldsymbol{A}) \right] = 0 \,. \tag{1.96}$$

The most general integration of this expression is of the form:

$$\frac{\partial \boldsymbol{A}}{\partial t} - \boldsymbol{u} \times (\nabla \times \boldsymbol{A}) + \eta \nabla \times (\nabla \times \boldsymbol{A}) = C \nabla \varphi \,, \tag{1.97}$$

where C is an arbitrary constant and the scalar function φ, arising from this "uncurling" of the induction equation has no effect on $\boldsymbol{B}$. It may contribute to the electric field $\boldsymbol{E}$, however, and so φ is conveniently regarded as the electrostatic potential. Now, a classical vector identity (see Appendix A) allows to develop the third term on the LHS as

$$\eta \nabla \times (\nabla \times A) = -\eta \nabla^2 A + \eta \nabla (\nabla \cdot A) \ . \tag{1.98}$$

Remember that the electrostatic potential φ is an arbitrary function, and can be picked at our convenience. Setting $C = \eta$ and picking the Coulomb gauge $\varphi = \nabla \cdot A$ simplifies the Ohmic dissipation term to $\eta \nabla^2 A$, so that the induction equation for the vector potential becomes:

$$\frac{\partial A}{\partial t} = u \times (\nabla \times A) + \eta \nabla^2 A \ . \tag{1.99}$$

Note that the Laplacian operator is here acting on a vector quantity, which, except for cartesian coordinates, differs from the Laplacian acting on individual vector components by the appearance of additional metric terms (see Appendix B).

1.12 Magnetic Helicity

In analogy with fluid helicity, one can define the *magnetic helicity* as

$$h_B = A \cdot B \ . \tag{1.100}$$

Consider now the variation of the total magnetic helicity ($\mathcal{H}_B$) in a co-moving fluid volume V; making judicious use of Eqs. (1.59), (1.97), and (1.55), a good deal of vector algebra eventually leads to the following evolution equation for ($\mathcal{H}_B$):

$$\frac{D}{Dt} \underbrace{\int_V A \cdot B \, dV}_{\mathcal{H}_B} = -2\mu_0 \eta \underbrace{\int_V J \cdot B \, dV}_{\mathcal{H}_J} \ , \tag{1.101}$$

where the integral on the RHS defines the total *current helicity* $\mathcal{H}_J$, which measure the topological linkage between magnetic fieldlines and electrical currents within the volume, much like the way in which the total magnetic helicity $\mathcal{H}_B$ measures the linkage of magnetic flux systems within V.

Equation (1.101) indicates that in the ideal MHD limit, magnetic helicity becomes a conserved quantity. This will turn out to pose a severe constraint on magnetic field amplification in astrophysical dynamos, an issue to which we will return in due time.

1.13 Force-Free Magnetic Fields

In many astrophysical systems, the magnetic field dominates the dynamics and energetics of the system. Left to itself, such a system would tend to evolve to a force-free state described by

1.13 Force-Free Magnetic Fields

$$F = J \times B = 0 \ . \tag{1.102}$$

Broadly speaking, this can be achieved in two physically distinct ways (excluding the trivial solution $B = 0$). The first is $J = 0$ throughout the system. Then Ampère's law becomes $\nabla \times B = 0$, which means that, as with the electric field in electrostatic, B can be expressed as the gradient of a potential. Such a magnetic field is called a *potential field*. Substitution into $\nabla \cdot B = 0$ then yields a Laplace-type problem:

$$B = \nabla \varphi \ , \quad \nabla^2 \varphi = 0 \quad \text{[Potential field]} \ . \tag{1.103}$$

Alternately, a system including a non-zero current density can still be force free, provided the currents flow everywhere parallel to the magnetic field, i.e.,

$$\nabla \times B = \alpha B \ , \tag{1.104}$$

where α need not necessarily be a constant, i.e., it can vary from one fieldline to another, vary in space, and even depend on the (local) value of B. Imagine now a situation where, in some domain (for example, the exterior of a star), we are provided with a boundary condition on B and the task is to construct a force-free field. Adopting the potential field Ansatz can lead to very different reconstructions than if we adopt instead Eq. (1.104), given that in the latter case one is free to specify any electric current distribution within the domain, as long as J remains parallel to B.

A very important result in this context is known as *Aly's theorem*; it states that in a semi-infinite domain with $B_\perp$ imposed at the boundary and $B \to 0$ as $x \to \infty$, the (unique) potential field solution satisfying the boundary conditions has a magnetic energy that is *lower* than any of the (multiple) solutions of Eq. (1.104) that satisfy the same boundary conditions, even with complete freedom to specify $\alpha(x)$ within the domain. This poses a strict limit to the amount of magnetic energy stored into a system that can actually be tapped into to power astrophysically interesting phenomena.

1.14 The Ultimate Origin of Astrophysical Magnetic Fields

1.14.1 Why B and not E ?

Pretty much anywhere we look in the known universe, there are magnetic fields of all strengths and shapes everywhere; but electric fields are conspicuously absents. Why is that? You might think, looking at Maxwell's equations (1.51)–(1.54) that E and B appear therein on apparently equal footing, leaving nothing to allow us to anticipate the observed astrophysical preponderance of magnetic fields over electrical fields. Moreover, one observer's magnetic field can be turned into another's electric field by a simple change of reference frame. So what's the deal here?

Well, for one thing if you use any sort of sensible "rest frame" for astronomical observation (Earth at rest; solar system at rest; Milky Way at rest; local group at rest; etc. ad infinitum) there is a lot of $\boldsymbol{B}$ around and precious little $\boldsymbol{E}$. The crucial difference between $\boldsymbol{E}$ and $\boldsymbol{B}$ in Maxwell's equations is not the fields themselves, nor the reference frame in which they are measured, but their *sources*. The Universe may be largely empty, but the fact is that it contains a whopping number of electrically charged particles of various sorts (free electrons, ionized atoms or molecules, photoelectrically charged dust grains, etc). If a large-scale electric field were suddenly to be turned on, all these charges will do the honorable thing, which is to separate along the electric field direction until the secondary electric field so produced cancels the externally applied electric field, at which point charge separation ceases. Moreover, the low densities of most astrophysical plasmas lead to very large mean-free paths for microscopic constituents, leading in turn to fairly good electrical conductivities and very short electrostatic relaxation times τ_e (see Eq. (1.71)), even when the ionisation fraction is quite low (such as in molecular clouds). In other words, astrophysical electric fields, if and whenever they appear, get shortcircuited mighty fast.

Not so with magnetic fields. For starters, as far as anyone can tell there are no magnetic monopoles out there (well, maybe just one, of primordial origin... more on this shortly), so shortcircuiting the magnetic field by monopole separation is out of the question. Magnetic fields, left to themselves, will simply decay as the electrical currents that support them (remember Ampère's law) suffer Ohmic dissipation. We already obtained a timescale for this process given by Eq. (1.63), and we already noted, on the basis of the compilation presented in Table 1.2, that this timescale is extremely large, often exceeding the age of the universe. Once magnetic fields are produced, by whatever means, they stick around for a long, long time. But when and how do they first appear? If we remain within the realm of MHD, then we immediately hit a Big Problem, arising from the linearity of the MHD induction equation (1.59): if $\boldsymbol{B} = 0$ at some time t_0 then $\boldsymbol{B} = 0$ at all subsequent times $t > t_0$, a problem that persists unabated as t_0 is pushed all the way back to the Big Bang. We need something else.

1.14.2 Monopoles and Batteries

In subsequent chapters we will see that astrophysical flows are actually quite apt at amplifying magnetic fields, so what we are after here is a very small *seed field* to start up the process. Cheap and easy explanations along the line of an original seed magnetic field being a primordial relic of the Big Bang need not concern us here. Nor is early-universe ferromagnetism a viable option, since permanent magnets require an externally-applied magnetic field to become magnetized in the first place.

1.14 The Ultimate Origin of Astrophysical Magnetic Fields

Interestingly, the two options that are currently deemed viable stand at the opposite ends of the physical exoticism scale: magnetic monopoles... and batteries.[8]

Already back in 1931 Paul Dirac (1902–1984) pointed out that there is nothing to prevent there being magnetic monopoles so long as the magnetic charge on a particle is some integer multiple of $g \equiv hc/(4\pi e) \approx 69e$, where h is Planck's constant, and e is the fundamental electric charge. With just *one* magnetic monopole in the universe we have our basic seed field. Some Grand Unified (field) Theories "predict" that very early in the formation of the universe a lot of $m_g \approx 10^{16}$ GeV/c^2 magnetic monopoles should be produced, to the point that an inflationary cosmological scenarios are needed to ensure that only a few such massive monopoles end up within each inflated subdomain.

In light of the fact that no one has yet seen a magnetic monopole, it would be wise to find a more pedestrian means to create seed magnetic fields, relying on basic physics that we know functions sensibly at least in our part of the universe. To this end, we return to our derivation of the induction equation (Sect. 1.3). Recall that one essential step toward MHD from Maxwell required stipulating Ohm's law, in the form of Eq. (1.57) for the laboratory frame of reference. Consider now the possibility of a "mechanically-driven" process of charge separation (i.e., not related to the presence of an electric field in any reference frame); Ohm's law then picks up an extra term:

$$\boldsymbol{J} = \sigma\left[\boldsymbol{E} + \boldsymbol{u} \times \boldsymbol{B}\right] + \boldsymbol{J}_{\text{mech}} . \qquad (1.105)$$

If we keep only the very first term on the RHS of Eq. (1.105), *and* drop the displacement current in Eq. (1.54), then we get back to the induction equation (1.59). If we avail ourselves of neither of these opportunities then we obtain instead:

$$\left\{1 + \frac{\eta}{c^2}\frac{\partial}{\partial t}\right\}\frac{\partial \boldsymbol{B}}{\partial t} = \nabla \times \left(\boldsymbol{u} \times \boldsymbol{B} - \eta\nabla \times \boldsymbol{B} + \eta\mu_0 \boldsymbol{J}_{\text{mech}}\right) . \qquad (1.106)$$

Notice that our only hope for creating $\boldsymbol{B}$ out of nothing (so to speak) is the $\boldsymbol{J}_{\text{mech}}$ term; retaining the displacement current gives us no advantage.

The $\boldsymbol{J}_{\text{mech}}$ term represents our ability to mechanically grab a hold of electric charges and force currents to flow; in other words, an electromotive force. In the dense interior of a conducting star, plasma kinetic theory permits one to write down a prescription for this "battery" contribution to the total electric current density as:

$$\boldsymbol{J}_{\text{mech}} = \frac{\sigma}{en_e}\left[\nabla p_e - \frac{1}{c}\boldsymbol{J} \times \boldsymbol{B}\right] , \qquad (1.107)$$

where p_e is the contribution of the electrons alone to the thermal pressure (see references in bibliography). For a completely ionized pure hydrogen plasma, p_e is just half of the total gas pressure, and $n_e = \varrho/m_p$, and so,

---

[8] This section is adapted from class notes written by Thomas J. Bogdan for the graduate class APAS7500 we co-taught in 1997 at CU Boulder.

$$\boldsymbol{J}_{\text{mech}} = \frac{\sigma m_p}{2e\varrho}\left[\nabla p - \frac{2}{c}\boldsymbol{J}\times\boldsymbol{B}\right]. \tag{1.108}$$

Now, the second term on the RHS of Eq. (1.108) does not do us any good since it carries a factor of $\boldsymbol{B}$, so the whole plan rests upon the first term generating a seed magnetic field. For a spherically symmetric star, we know from hydrostatic equilibrium that $\nabla\Phi = (\nabla p)/\varrho$, and so the product $\eta \boldsymbol{J}_{\text{mech}} \propto \nabla\Phi$. This will not work because of the curl operator on the RHS of Eq. (1.106) will yield zero upon acting on $\boldsymbol{J}_{\text{mech}}$ since (∇p) is a gradient of a scalar function. How can we get around this constraint? A viable possibility is rotation. If a star is rotating, then there is a centrifugal force per unit density of $s\Omega^2 \hat{\boldsymbol{e}}_s$ which adds to $\nabla\Phi$ and which leads to the generation of a seed magnetic field. This process of the centrifugal force driving a flow of electrons relative to the ions was first pointed out by Ludwig Biermann (1907–1986) and is now called the *Biermann battery*.

In fact *any* process that can produce a relative motion between the ions and electrons is a potential battery mechanism, and a possible candidate for creating seed magnetic fields. For example, consider a rotating proto-galaxy, where the outer portions of the proto-galaxy move at a speed $U = R\Omega$ relative to the frame in which the microwave background is isotropic. The Thomson scattering of the microwave photons by the electrons results in the so-called *Compton drag effect*, which causes the electrons to counter-rotate with respect to the ions. The net result is an azimuthal current which generates a dipole-like magnetic field.

Of course, if you bother to put typical numbers in these various examples you will find that you don't really generate very *much* magnetic field. But generating a lot of field is not the point, that can be done via magnetic flux conservation in a collapsing protostellar cloud, or, as we shall see in due time, via the $\boldsymbol{u} \times \boldsymbol{B}$ term in our MHD induction equation. The basic idea to take away from this section is that invoking weird, unproven physics to get away from $\boldsymbol{B} = 0$ is not necessary.

1.15 The Astrophysical Dynamo Problem(s)

Before moving on with astrophysical dynamos, it will prove instructive to first consider the following example of a simple laboratory dynamo, which illustrates nicely how the idea of amplifying magnetic field by bodily moving electrical charges across a magnetic field is not so mysterious as one may initially think.

1.15.1 A Simple Dynamo

One of the many practical inventions of Michael Faraday (1791–1867) was a DC electric current generator based on the rotation of a conducting metallic disk threaded by an external magnetic field. Figure 1.5a illustrates the basic design: a circular disk

1.15 The Astrophysical Dynamo Problem(s)

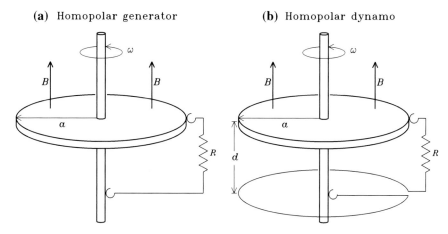

Fig. 1.5 A homopolar generator (**a**) versus a homopolar dynamo (**b**). An external magnetic field B is applied across a rotating conducting disk, producing an electromotive force that drives a radial current, a wire connecting the edge of the disk to the axle, forming a circuit of resistance R. The only difference between the two electro-mechanical devices illustrated here is that in the latter case, the wire completing the circuit by connecting on the axle is wrapped into a loop in a plane parallel to the disk, so that a secondary vertical magnetic field is produced (see text).

of radius a mounted on an axle, rotating at angular velocity ω through the agency of some external mechanical force (e.g., Faraday turning a crank). A vertical magnetic field is imposed across the disk. Electrical charges in the disk will feel the usual Lorentz force $\boldsymbol{F} = q\boldsymbol{v} \times \boldsymbol{B}$ where, (initially) $\boldsymbol{v}$ is just the motion imposed by the rotation of the disk. Working in cylindrical coordinates (s, ϕ, z) one can write

$$\boldsymbol{v} = (\omega s)\hat{\boldsymbol{e}}_\phi, \qquad \boldsymbol{B} = B_0 \hat{\boldsymbol{e}}_z, \tag{1.109}$$

so that

$$\boldsymbol{F} = (q\omega s B_0)\hat{\boldsymbol{e}}_s. \tag{1.110}$$

Now consider the circuit formed by connecting the edge of the disk to the base of the axle via frictionless sliding contacts. With the lower part of the circuit away from the imposed magnetic field, the only portion of the circuit where the magnetic force acts on the charges is within the disk, amounting to an electromotive force

$$\mathcal{E} = \oint_{\text{circuit}} \left(\frac{\boldsymbol{F}}{q}\right) \cdot d\boldsymbol{\ell} = \int_0^a \omega B_0 s \, ds = \frac{\omega B_0 a^2}{2}. \tag{1.111}$$

Neglecting for the time being the self-inductance of the circuit, the current flowing through the resistor is simply given by $I = \mathcal{E}/R$. This device is called a *homopolar generator*.

There is a subtle modification to this setup that can turn this generator into a *homopolar dynamo*, namely a device that converts mechanical energy into self-amplifying electrical currents and magnetic fields. Instead of simply connecting the resistor straight to the axle as on Fig. 1.5a, the wire is wrapped around the axle in a loop lying in a plane parallel to the disk, and then connected to the axle, as shown on Fig. 1.5b. Use your right-hand rule to convince yourself that this current loop will now produce a secondary magnetic field B_* that will superpose itself on the external field B_0. The magnetic flux through the disk associated with this secondary field will be proportional to the current flowing in the wire loop, the proportionality constant being defined as the inductance (M):

$$MI = \Phi = \pi a^2 B_* , \tag{1.112}$$

where the second equality comes from assuming that the secondary field is vertical and constant across the disk; but what really matters here is that $B_* \propto I$ since the geometry is fixed. We now write an equation for the electrical current, this time taking into consideration the counter-electromotive force associated with self-inductance of the circuit since $\boldsymbol{B}$, and therefore also I, are time-varying:

$$\mathcal{E} - L\frac{dI}{dt} = RI , \tag{1.113}$$

where L is the coefficient of self-inductance. Substituting Eqs. (1.111) and (1.112) into this expression, leads to

$$L\frac{dI}{dt} = \frac{\omega a^2}{2}\left(B_0 + \frac{MI}{\pi a^2}\right) - RI , \tag{1.114}$$

indicating that the current—and thus the magnetic field—will grow provided that initially,

$$\frac{\omega a^2 B_0}{2} > RI , \tag{1.115}$$

which it certainly will at first since $I = 0$ at $t = 0$. There will eventually come a time (t_*) when the secondary magnetic field will be comparable in strength to the externally applied field B_0, at which point we may as well "disconnect" B_0; Eq. (1.114) then becomes

$$L\frac{dI}{dt} = \left(\frac{\omega M}{2\pi} - R\right)I , \tag{1.116}$$

which integrates to

$$I(t) = I(t_*)\exp\left[\frac{1}{L}\left(\frac{\omega M}{2\pi} - R\right)t\right] , \tag{1.117}$$

1.15 The Astrophysical Dynamo Problem(s)

indicating that the current—and magnetic field—will grow provided the externally-imposed angular velocity exceeds a critical value:

$$\omega > \omega_c = \frac{2\pi R}{M} . \qquad (1.118)$$

This is not a (dreaded) case of perpetual motion, or creating energy out of nothing, or anything like that. The energy content of the growing magnetic field ultimately comes from the biceps of the poor experimenter working ever harder and harder to turn the crank and keep the angular velocity ω at a constant value.

There are many features of this dynamo system worth noting, and which all find their equivalent in the MHD dynamos to be studied in chapters to follow:

1. There exist a critical angular velocity that must be reached for the self-inductance to beat Ohmic dissipation in the resistor, leading to an exponential growth of the magnetic field; below this critical value, the field decays away exponentially once the initial field B_0 is removed;
2. Not all circuits connecting the edge of the disk to the axle will operate in this way; if we wrap the wire the other way around the axle, the magnetic field produced by the loop will *oppose* the applied field;
3. The externally applied magnetic field B_0 is only needed as a *seed field* to initiate the amplification process;
4. The homopolar dynamo is really nothing more than a device turning mechanical energy into electromagnetic energy, more specifically magnetic energy.

1.15.2 The Challenges

Copper disks and sliding contacts being a rather sparse commodity in the universe, we must now figure out to apply the general idea of a dynamo to astrophysical fluids. In the MHD limit, our hope lies evidently with the induction term $\nabla \times (\boldsymbol{u} \times \boldsymbol{B})$ in the induction equation (1.59).

In its simplest form, the *dynamo problem* consists in finding a flow field $\boldsymbol{u}$ that can sustain a magnetic field against Ohmic dissipation. We will encounter in the following chapter flows that can amplify a magnetic field during a transient time interval, after which $\boldsymbol{B}$ decays again. So we tighten our definition of the dynamo problem by demanding that a flow be a dynamo if it can lead to $\mathcal{E}_B > 0$ for times much larger than all relevant advective and diffusive timescales of the problem. To make things even harder, we'll add the additional condition that no electromagnetic energy be supplied across the domain boundaries, i.e., $\boldsymbol{S} \cdot \boldsymbol{n} = 0$ in Eq. (1.87).

We must distinguish the *kinematic dynamo problem*, where the flow field $\boldsymbol{u}$ is considered given a priori and constructed without any regards for its underlying dynamics, from what can only be called (for lack of a generally agreed-upon terminology) the *full dynamo problem*, in which the flow $\boldsymbol{u}$ results from a solution of the full set of MHD equations (Sect. 1.7), including the backreaction of the magnetic field on the flow via the Lorentz force term $\boldsymbol{J} \times \boldsymbol{B}$ on the RHS of the Navier–Stokes equation. The kinematic regime carries the immense practical advantage that the induction equation then becomes truly linear in $\boldsymbol{B}$, and the dynamo problem reduces to finding a (smooth) flow field $\boldsymbol{u}$ that has the requisite topological properties to lead to field amplification. In the following chapters we will concentrate mostly on this kinematic regime, but will occasionally touch upon the much more difficult dynamical problem.

The *solar dynamo problem* can be tackled either in kinematic or fully dynamical form. The aim there is to reproduce observed spatiotemporal patterns of solar (and stellar) magnetic field evolution, including things like cyclic polarity reversals, equatorward migration of activity belts, relative strengths and phase relationships between poloidal and toroidal component, etc. This will prove to be a very tall order. Yet, from solar irradiance variations and their possible influence on Earth's climate, space weather prediction, and the understanding of stellar magnetic fields, it all begins with the solar cycle. Keep this in mind as we now start to dig into the mathematical and physical intricacies of magnetic field generation in electrically conducting fluids. We'll seem to venture pretty far away from the sun and stars at times, but stick to it and you'll see it all fitting together at the end. And now, into the abyss...

Bibliography

There are a great many textbooks available on classical hydrodynamics. My own top-three personal favorites are

Tritton, D. J.: 1988, *Physical Fluid Dynamics*, 2nd edn., Oxford University Press
Acheson, D. J.: 1990, *Elementary Fluid Dynamics*, Clarendon Press
Landau, L., & Lifschitz, E.: 1959, *Fluid Mechanics*, Pergamon Press

If you need a refresher on undergraduate electromagnetism, I would recommend

Griffiths, D. J.: 1999, *Introduction to Electrodynamics*, 3rd edn., Prentice-Hall

At the graduate level, the standard reference remains

Jackson, J. D.: 1975, *Classical Electrodynamics*, John Wiley & Sons

who does devote a chapter to magnetohydrodynamics, including a discussion of magnetic wave modes. My personal favorite on magnetohydrodynamics is:

Davidson, P. A.: 2001, *An Introduction to Magnetohydrodynamics*, Cambridge University Press

Sects. 1.5 and 1.10 are strongly inspired by Davidson's own presentation of the subject. He also presents an illuminating proof of the kinematic theorem embodied in Eq. (1.95). The following textbook is also well worth consulting:

Goedbloed, H., & Poedts, S.: 2004, *Principles of Magnetohydrodynamics*, Cambridge University Press

These authors put greater emphasis on MHD waves, shocks, and on the intersection of MHD and plasma physics. For those seeking even more focus on plasma physics aspects, I would recommend:

Bibliography

Kulsrud, R. M.: 2005, *Plasma Physics for Astrophysics*, Princeton University Press

Also noteworthy in the general astrophysical context:

Shu, F. H.: 1992, *The Physics of Astrophysics*, Vol. I and II, University Science Books
Choudhuri, A. R.: 1998, *The Physics of Fluids and Plasmas*, Cambridge University Press

On Aly's theorem, see

Aly, J. J.: 1991, *How much energy can be stored in a three-dimensional force-free magnetic field?*, Astrophys. J. Lett., **375**, L61–L64
Low, B. C., & Smith, D.F.: 1993, *The free energies of partially open coronal magnetic fields*, Astrophys. J., **410**, 412–425

but brace yourself for some serious math. Many ambitious monographs have been written on the general topic of astrophysical magnetic fields. My personal "top-three" selection is:

Parker, E. N.: 1979, *Cosmical Magnetic Fields: Their Origin and their Activity*, Clarendon Press
Mestel, L.: 1999, *Stellar Magnetism*, Clarendon Press
Rüdiger, G., & Hollerbach, R.: 2004, *The Magnetic Universe*, Wiley-VCH

Parker's book is unfortunately out of print, and Rüdiger & Hollerbach's outrageously priced. The Mestel book was issued in paperback in 2003, but is still quite pricey.

For some "light" reading on magnetic monopoles in field theory and astrophysics, try,

Dirac, P. A. M.: 1931, *Quantised singularities in the electromagnetic field*, Proc. Roy. Soc. London Ser. A, **133**, 60–72
Parker, E. N.: 1970, *The origin of magnetic fields*, Astrophys. J., **160**, 383–404
Cabrera, B.: 1982, *First results from a superconductive detector for moving magnetic monopoles*, Phys. Rev. Lett., **48**, 1378–1381
Kolb, E. W., & Turner, M. S.: 1990, *The Early Universe*, Addison-Wesley, Sect. 7.6

and references therein. For more on Biermann's battery, see, additional to Chap. 13 in the Kulsrud book cited earlier,

Biermann, L.: 1950, *Über den Ursprung der Magnetfelder auf Sternen und im interstellaren Raum (mit einem Anhang von A. Schlüter)*, Zeitschrift Naturforschung Teil A, 5, 65
Roxburgh, I. W.: 1966, *On stellar rotation, III. Thermally generated magnetic fields*, Mon. Not. Roy. Astron. Soc., **132**, 201–215
Chakrabarti, S. K., Rosner, R., & Vainshtein, S. I: 1994, *Possible role of massive black holes in the generation of galactic magnetic fields*, Nature, **368**, 434–436

Chapter 2
Decay and Amplification of Magnetic Fields

> *It's not whether a thing is hard to understand.*
> *It's whether, once understood, it makes any sense.*
>
> Hans Zinsser
> Rats, Lice and History (1934)

We now begin our long modelling journey towards astrophysical dynamos. This chapter concentrate for the most part on a series of (relatively) simple model problems illustrating the myriad of manners in which a flow and a magnetic field can interact. We first consider the purely resistive decay of magnetic fields (Sect. 2.1), then examine various circumstances under which stretching and shearing by a flow can amplify a magnetic field (Sect. 2.2). This is followed by a deeper look at some important subtleties of these processes in the context of some (relatively) simple 2D flows (Sect. 2.3). We then move on to the so-called anti-dynamo theorems (Sect. 2.4), which will shed light on results from previous sections and indicate the way towards dynamo action, which we will finally encounter in Sects. 2.5 and 2.6.

Some of the material contained in this chapter may feel pretty far removed from the realm of astrophysics at times, but please do stick to it because the physical insight (hopefully) developed in the following sections will prove essential to pretty much everything that will come next.

2.1 Resistive Decays of Magnetic Fields

Before we try to come up with flows leading to field amplification and dynamo action, we better understand the enemy, namely magnetic field decay by Ohmic dissipation. Consequently, and with the sun and stars in mind, we first consider the evolution of magnetic fields in a sphere (radius R) of electrically conducting fluid, in the absence of any fluid motion (or, more generally, in the $R_m \ll 1$ limit). The induction equation then reduces to

$$\frac{\partial \boldsymbol{B}}{\partial t} = -\nabla \times (\eta \nabla \times \boldsymbol{B}) = \eta \nabla^2 \boldsymbol{B} - (\nabla \eta) \times (\nabla \times \boldsymbol{B}) \,. \tag{2.1}$$

Were it not that we are dealing here with a vector—as opposed to scalar—quantity, for constant η this would look just like a simple heat diffusion equation, with η playing the role of thermal diffusivity. Our derivation of the magnetic energy equation (1.87) already indicates that under such circumstances, the field can only decay. Back in Chap. 1 we already obtained an order-of-magnitude estimate for the timescale $\tau_\eta \sim \ell^2/\eta$ over which a magnetic field $\boldsymbol{B}$ with typical length scale ℓ can be expected to resistively decay, which in the case of the stellar interiors ended up at $\sim 10^{10}$ yr, i.e., about the main-sequence lifetime of the sun. Let's now validate this estimate by securing formal solutions to the diffusive decay problem.

2.1.1 Axisymmetric Magnetic Fields

Without any significant loss of generality, we can focus on *axisymmetric* magnetic fields, i.e., fields showing symmetry with respect to an axis, usually rotational. Working in spherical polar coordinates (r, θ, ϕ) with the polar axis coinciding with the field's symmetry axis, the most general axisymmetric (now meaning $\partial/\partial\phi = 0$) magnetic field can be written as:

$$\boxed{\boldsymbol{B}(r, \theta, t) = \nabla \times (A(r, \theta, t)\hat{\boldsymbol{e}}_\phi) + B(r, \theta, t)\hat{\boldsymbol{e}}_\phi} \,. \tag{2.2}$$

Here the vector potential component A defines the *poloidal* component of the magnetic field, i.e., the component contained in meridional (r, θ) planes. The azimuthal component B is often called the *toroidal field*. Equation (2.2) satisfies the constraint $\nabla \cdot \boldsymbol{B} = 0$ by construction, and another great advantage of this mixed representation is that the MHD induction equation for the vector $\boldsymbol{B}$ can be separated in two equations for the scalar components A and B. The trick is the following: substitution of Eq. (2.2) into (2.1) leads to a series of (vector) terms, some oriented in the (toroidal) $\hat{\boldsymbol{e}}_\phi$-direction, others perpendicularly, in the (poloidal) meridional plane. The original, full induction equation can only be satisfied if the two sub-equations defined by each sets of orthogonal terms are individually satisfied, thus defining two separate evolution equations for A and B.

In the case of pure diffusive decay, and for a magnetic diffusivity η depending at worst only on r, this poloidal/toroidal separation leads to:

$$\frac{\partial A}{\partial t} = \eta \left(\nabla^2 - \frac{1}{\varpi^2} \right) A \,, \tag{2.3}$$

$$\frac{\partial B}{\partial t} = \eta \left(\nabla^2 - \frac{1}{\varpi^2} \right) B + \frac{1}{\varpi} \frac{\partial \eta}{\partial r} \frac{\partial (\varpi B)}{\partial r} \,, \tag{2.4}$$

where $\varpi = r\sin\theta$. These are still diffusion-like PDEs, now fully decoupled from one another. In the "exterior" $r > R$ there is only vacuum, which implies vanishing electric currents. In practice we will need to match whatever solution we compute in $r < R$ to a current-free solution in $r > R$; such a solution must satisfy

$$\mu_0 \mathbf{J} = \nabla \times \mathbf{B} = 0 \,. \tag{2.5}$$

For an axisymmetric system, Eq. (2.5) translates into the requirement that

$$\left(\nabla^2 - \frac{1}{\varpi^2}\right) A(r,\theta,t) = 0\,, \qquad r > R\,, \tag{2.6}$$

$$B(r,\theta,t) = 0\,, \qquad r > R\,. \tag{2.7}$$

Solutions to Eq. (2.6) have the general form

$$A(r,\theta,t) = \sum_{l=1}^{\infty} a_l \left(\frac{R}{r}\right)^{l+1} Y_{l0}(\cos\theta) \qquad r > R\,, \tag{2.8}$$

where the Y_{l0} are the usual spherical harmonics of $m = 0$ azimuthal order, and l is a positive integer, modes with negative l being discarded to ensure proper behavior as $r \to \infty$.

2.1.2 Poloidal Field Decay

Let us now seek specific solutions for a few situations of solar/stellar interest.[1] The first point to note is that the coefficients that appear in Eqs. (2.3)–(2.4) have no explicit dependence on time; provided that the magnetic diffusivity η is at worst only a function of r, it is then profitable to seek a separable solution of the form:

$$e^{-\lambda t} f_\lambda(r) Y_{lm}(\theta,\phi) \,, \tag{2.9}$$

where the Y_{lm} are again the spherical harmonics, the natural functional basis for modal development on a spherical surface. Substitution of this Ansatz into Eq. (2.3) with $m = 0$ in view of axisymmetry, yields the ODE:

$$\left[\frac{1}{r^2}\frac{\mathrm{d}}{\mathrm{d}r}r^2\frac{\mathrm{d}}{\mathrm{d}r} - \frac{l(l+1)}{r^2} + \frac{\lambda}{\eta(r)}\right] f_\lambda(r) = 0 \,. \tag{2.10}$$

---

[1] This and the following subsection are to a large extent adapted from class notes written by Thomas J. Bogdan for the graduate class APAS7500 we co-taught in 1997 at CU Boulder.

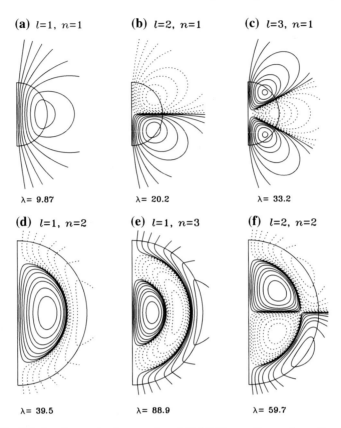

Fig. 2.1 Six diffusive eigenmodes for a purely poloidal field pervading a sphere of constant magnetic diffusivity embedded in vacuum. The *top row* shows the three fundamental ($n = 1$) diffusive eigenmodes with smallest eigenvalues, i.e., largest decay times. They correspond to the well-known dipolar, quadrupolar, and hexapolar modes ($l = 1, 2$ and 3). The *bottom row* shows a few eigenmodes of higher radial overtones. Poloidal fieldlines are shown in a meridional plane, and the eigenvalues are given in units of the inverse diffusion time ($\tau^{-1} \sim \eta/R^2$).

Assume now that the magnetic diffusivity η is constant; the spherical Bessel functions $j(kr)$, with $k^2 = \lambda/\eta$, are then the appropriate solution. The decay rate, λ, is then determined by the above 1D eigenvalue problem, along with some boundary conditions at the surface of the sphere, which turns out to depend on the vector character of the decaying magnetic field.

We first consider the decay of a purely poloidal field, i.e., $f_\lambda(r)$ is taken to describe the radial dependency of the toroidal vector potential component $A(r, \theta, t)$. Both the interior solution and outer potential field solution carry the Y_{l0} angular dependency, so continuity of A at $r = R$ demands that

2.1 Resistive Decays of Magnetic Fields

$$f_\lambda(r) = \begin{cases} j_l(kr) & r < R, \\ j_l(kR)\left(\frac{R}{r}\right)^{l+1} & r > R. \end{cases} \quad (2.11)$$

The continuity of the radial derivative at $r = R$, necessary for the continuity of the latitudinal component of the magnetic field, then requires

$$kRj_l'(kR) + (l+1)j_l(kR) = kRj_{l-1}(kR) = 0, \quad (2.12)$$

which means that the decay rate of a poloidal magnetic field is determined by the zeros of a spherical Bessel function. An $l = 1$ dipole calls for the positive zeros of $j_0(x) = \sin x / x$:

$$\lambda_n = \frac{\eta \pi^2 n^2}{R^2} \quad \text{for } l = 1, \; n = 1, 2, 3, \dots . \quad (2.13)$$

Notice the many possible *overtones* associated with $n \geq 2$. These decay more rapidly than the fundamental ($n = 1$), since the radial eigenfunctions possess $n - 1$ field reversals. For such overtones, the effective length scale to be used in the decay-time estimate is roughly the radial distance between the field reversals, or $\approx R/n$.

Figure 2.1 (top row) shows the first three fundamental ($n = 1$) modes of angular degrees $l = 1, 2, 3$, corresponding to dipolar, quadrupolar, and hexapolar magnetic fields, as well as a few higher overtones for $l = 1, 2$ (bottom row). The decay time estimate provided by Eq. (1.63) turns out to be too large by a factor $\pi^2 \approx 10$, for a sun with constant diffusivity. Still not so bad for a pure order-of-magnitude estimate!

2.1.3 Toroidal Field Decay

Computing the decay rate of a purely toroidal magnetic field follows the same basic logic. We now require $B = 0$ at $r = R$, and the decay rate ends up related to the zero of a spherical Bessel function—only of index l rather than $l - 1$ as was found for the decay of the poloidal field. Hence, a dipole ($l = 1$) toroidal magnetic field decays at precisely the same rate as a quadrupole ($l = 2$) poloidal magnetic field (still for constant diffusivity). Sneaking a peak in a handbook of special functions soon reveals that the decay rate of a $l = 1$ toroidal field follows from the transcendental equation:

$$\tan kR = kR . \quad (2.14)$$

The smallest non-zero solution of this equation gives,

$$\lambda_1 = \frac{\eta(4.493409\dots)^2}{R^2}, \quad l = 1 \text{ toroidal and } l = 2 \text{ poloidal} . \quad (2.15)$$

As with a purely poloidal field, higher radial overtones decay proportionally faster.

2.1.4 Results for a Magnetic Diffusivity Varying with Depth

We end this section by a brief examination of the diffusive decay of large-scale poloidal magnetic fields in the solar interior. The primary complication centers on the magnetic diffusivity, which is no longer constant throughout the domain, and turns out to be rather difficult to compute from first principles. To begin with, the depth variations of the temperature and density in a solar model causes the magnetic diffusivity to increase from about $10^{-2}\,\mathrm{m^2\,s^{-1}}$ in the central core to $\sim 1\,\mathrm{m^2\,s^{-1}}$ at the core–envelope interface. This already substantial variation is however dwarfed by the much larger increase in the net magnetic diffusivity expected in the turbulent environment of the convective envelope. We will look into this in some detail in Chap. 3, but for the time being let us simply take for granted that η is much larger in the envelope than in the core.

In order to examine the consequences of a strongly depth-dependent magnetic diffusivity for the diffusive eigenmodes, we consider a simplified situation whereby η assumes a constant value η_c in the core, a constant value η_e ($\gg \eta_c$) in the envelope, the transition occurring smoothly across a thin spherical layer coinciding with the core–envelope interface. Mathematically, such a variation can be expressed as

$$\eta(r) = \eta_c + \frac{\eta_e - \eta_c}{2}\left[1 + \mathrm{erf}\left(\frac{r - r_c}{w}\right)\right], \tag{2.16}$$

where $\mathrm{erf}(x)$ is the error function, r_c is the radius of the core–envelope interface, and w is the half-width of the transition layer.

We are still facing the 1D eigenvalue problem presented by Eq. (2.10)! Expressing time in units of the diffusion time R^2/η_e based on the envelope diffusivity, we seek numerical solutions, subjected to the boundary conditions $f_\lambda(0) = 0$ and smooth matching to a potential field solution in $r/R > 1$, with the diffusivity ratio $\Delta\eta = \eta_c/\eta_e$ as a parameter of the model. Since we can make a reasonable guess at the first few eigenvalues on the basis of the diffusion time and adopted values of l and η_c ($\sim \pi^2 nl\Delta\eta$, for l and n not too large), a (relatively) simple technique such as inverse iteration is well-suited to secure both eigenvalues and eigenfunctions for the problem.

Figure 2.2 shows the radial eigenfunctions for the slowest decaying poloidal eigenmodes ($l = 1, n = 1$), with $r_c/R = 0.7$, $w/R = 0.05$ in Eq. (2.16) and diffusivity contrasts $\Delta\eta = 1$ (constant diffusivity), 10^{-1} and 10^{-3}. The corresponding eigenvalues, in units of R^2/η_e, are $\lambda = 9.87, 2.14$ and 0.028. Clearly, the (global) decay time is regulated by the region of *smallest* diffusivity, since λ scales approximately as $(\Delta\eta)^{-1}$. Notice also how the eigenmodes are increasingly concentrated in the core region ($r/R \lesssim 0.7$) as $\Delta\eta$ decreases, i.e., they are "expelled" from the convective envelope.

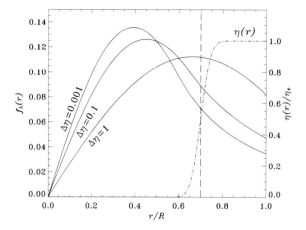

Fig. 2.2 Radial eigenfunctions for the slowest decaying ($\ell = 1$) poloidal eigenmodes ($l = 1$, $n = 1$) in a sphere embedded in a vacuum. The diffusivity is computed using Eq. (2.16) with $r_c/R = 0.7$, $w/R = 0.05$, and for three values of the core-to-envelope diffusivity ratio ($\Delta\eta$). The eigenvalues, in units of η_e/R^2, are $\lambda = 9.87$, 2.14 and 0.028 for $\Delta\eta = 1$, 0.1, and 10^{-3}, respectively. The diffusivity profile for $\Delta\eta = 10^{-3}$ is also plotted (*dash-dotted line*). The *vertical dashed line* indicates the location of the core–envelope interface.

2.1.5 Fossil Stellar Magnetic Fields

The marked decrease of the diffusive decay time with increasing angular and radial degrees of the eigenmodes is a noteworthy result. It means that left to decay long enough, any arbitrarily complex magnetic field in the sun or stars will eventually end up looking dipolar. Conversely, a fluid flow acting as a dynamo in a sphere and trying to "beat" Ohmic dissipation can be expected to preferentially produce a magnetic field approximating diffusive eigenmodes of low angular and radial degrees (or some combination thereof), since these are the least sensitive to Ohmic dissipation.

There exists classes of early-type main-sequence stars, i.e., stars hotter and more luminous than the sun and without deep convective envelope, that are believed to contain strong, large-scale fossil magnetic fields left over from their contraction toward the main-sequence. The chemically peculiar Ap/Bp stars are the best studied class of such objects. Reconstruction of their surface magnetic field distribution suggests almost invariably that the fields are dominated by a large-scale dipole-like component, as one would have expected from the preceding discussion if the observed magnetic fields have been diffusively decaying for tens or hundreds of millions of years. It is indeed quite striking that the highest strengths of large-scale magnetic fields in main-sequence stars (a few T in Ap stars), in white dwarfs ($\sim 10^5$ T) and in the most strongly magnetized neutron stars ($\sim 10^{11}$ T) all amount to similar total surface magnetic fluxes, $\sim 10^{19}$ Wb, lending support to the idea that these high field strengths can be understood from simple flux-freezing arguments (Sect. 1.10), with field amplification resulting directly from magnetic flux conservation as the star shrinks to form a compact object. We will revisit the origin of A-star magnetic fields in Chap. 5.

2.2 Magnetic Field Amplification by Stretching and Shearing

Having thus investigated in some details the resistive decay of magnetic field, we turn to the other physical mechanism embodied in Eq. (1.59): growth of the magnetic field in response to the inductive action of a flow u. We first take a quick look at field amplification in a few idealized model flows, and then move on to a specific example involving a "real" astrophysical flow.

2.2.1 Hydrodynamical Stretching and Field Amplification

Let's revert for a moment to the ideal MHD case ($\eta = 0$). The induction equation can then be expressed as

$$\left(\frac{\partial}{\partial t} + u \cdot \nabla\right) B = B \cdot \nabla u , \qquad (2.17)$$

where it was further assumed that the flow is incompressible ($\nabla \cdot u = 0$). The LHS of Eq. (2.17) is the Lagrangian derivative of B, expressing the time rate of change of B in a fluid element moving with the flow. The RHS expresses the fact that this rate of change is proportional to the local *shear* in the flow field. Shearing has the effect of *stretching* magnetic fieldlines, which is what leads to magnetic field amplification.

As a simple example, consider on Fig. 2.3 a cylindrical fluid element of length L_1, threaded by a constant magnetic field of strength B_1 oriented parallel to the axis of the cylinder. In MHD, such a magnetic field could be sustained by an azimuthal current concentrated in a thin sheet coinciding with the outer boundary of the cylinder, giving a solenoid-like current+field system. Assume now that this magnetic "flux tube" is embedded in a perfectly conducting incompressible fluid and subjected to a stretching motion ($\partial u_z/\partial z > 0$) along its central axis such that its length increases to L_2. Mass conservation demands that

$$\frac{R_2}{R_1} = \sqrt{\frac{L_1}{L_2}} . \qquad (2.18)$$

Conservation of the magnetic flux ($= \pi R^2 B$), in turn, leads to

2.2 Magnetic Field Amplification by Stretching and Shearing

Fig. 2.3 Stretching of a magnetized cylindrical fluid element by a diverging flow. The magnetic field (*fieldlines in gray*) is horizontal within the tube, has a strength B_1 originally, and B_2 after stretching. In the flux-freezing limit, mass conservation within the tube requires its radius to decrease, which in turn leads to field amplification (see text).

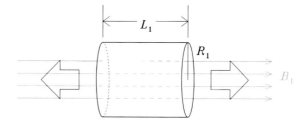
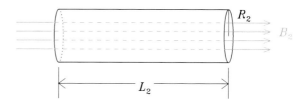

$$\frac{B_2}{B_1} = \frac{L_2}{L_1}, \qquad (2.19)$$

i.e., the field strength is amplified in direct proportion to the level of stretching.

This almost trivial result is in fact at the very heart of *any* magnetic field amplification in the magnetohydrodynamical context, and illustrates two crucial aspects of the mechanism: first, this works only if the fieldlines are frozen into the fluid, i.e., in the high-R_m regime. Second, mass conservation plays an essential role here; the stretching motion along the tube axis *must* be accompanied by a compressing fluid motion perpendicular to the axis if mass conservation is to be satisfied. It is this latter compressive motion, occurring perpendicular to the magnetic fieldlines forming the flux tube, that is ultimately responsible for field amplification; the horizontal fluid motion occurs parallel to the magnetic fieldline, and so cannot in itself have any inductive effect as per Eq. (1.59). This becomes evident upon considering the transfer of energy in this magnetized fluid system. With the electrical current sustaining the magnetic field concentrated in a thin cylindrical sheet bounding the tube, the field is force-free everywhere except at the surface of the tube, where the Lorentz force points radially outwards. It is the work done by the flow against this force which transfers energy from the flow to the magnetic field, and ultimately ends up in magnetic energy (viz. Eq. (1.87)).

The challenge, of course, is to realize this idealized scenario in practice, i.e., to find a flow which achieves the effect illustrated on Fig. 2.3. This, it turns out, is much simpler than one might expect! Working in cylindrical coordinates (s, ϕ, z), consider the following incompressible flow:

$$u_s(s) = \frac{\alpha s}{2}, \qquad u_\phi = 0, \qquad u_z(z) = -\alpha z, \qquad (2.20)$$

Fig. 2.4 Streamlines of the stagnation point flow defined by Eq. (2.20), plotted in a constant-ϕ plane. The flow is rotationally invariant about the symmetry (z) axis, indicated by the *dotted line*, and the stagnation point (*solid dot*) is located at the origin of the cylindrical coordinate system. A thin magnetic flux tube located in the $z = 0$ plane and crossing the origin, as shown, will be subjected to the stretching motion illustrated in cartoon form on Fig. 2.3.

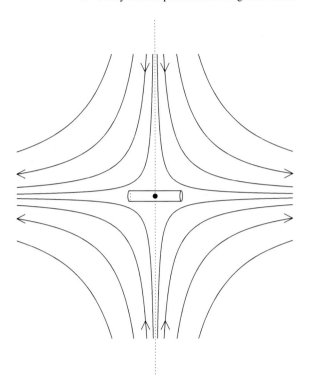

with $\alpha > 0$. This describes a flow converging towards the $z = 0$ plane along the z-axis, and diverging radially away from the origin within the $z = 0$ plane. Clearly, $u = 0$ at the origin $(0, \phi, 0)$ of the cylindrical coordinate system. This is called a *stagnation point*, and its presence is vital to the inductive amplification of the magnetic field. Now place a thin, straight magnetic flux tube in the $z = 0$ plane, and enclosing the stagnation point, as shown on Fig. 2.4. You can easily verify that you will get exactly the type of stretching effect illustrated in cartoon form on Fig. 2.3.

2.2.2 The Vainshtein & Zeldovich Flux Rope Dynamo

So, the linear stretching of a flux tube amplifies the magnetic field, but the magnetic flux remains constant by the very nature of the amplification mechanism. Nonetheless, this idea actually forms the basis of a dynamo that can increase both the magnetic field strength and flux. S. Vainshtein and Ya. B. Zeldovich have proposed one of the first and justly celebrated "cartoon" model for this idea, as illustrated on Fig. 2.5. The steps are the following:

1. A circular rope of magnetic field is stretched to twice its length ($a \to b$). As we just learned, this doubles the magnetic field strength while conserving the flux;

2.2 Magnetic Field Amplification by Stretching and Shearing

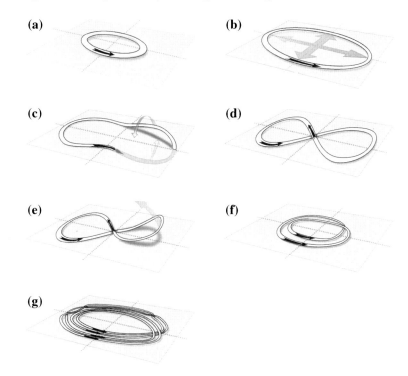

Fig. 2.5 Cartoon of the Stretch–Twist–Fold flux rope dynamo of Vainshtein & Zeldovich. A circular flux rope is (**a** → **b**) stretched, (**c** → **d**) twisted, and (**e** → **f**) folded. Diagram **g** shows the resulting structure after another such sequence acting on (**f**). Diagram produced by D. Passos.

2. The rope is twisted by half a turn ($c \to d$);
3. One half of the rope is folded over the other half in such a way as to align the magnetic field of each half ($e \to f$); this now doubles the magnetic flux through any plane crossed by the stacked loops.

This is quite remarkable; the so-called *stretch–twist–fold* sequence (hereafter STF) illustrated on Fig. 2.5 first doubles the field strength while conserving the magnetic flux of the original rope, because the tube's cross-section ($\propto R^2$) varies as the inverse of its length (as per Eq. 2.18), then folding doubles the magnetic flux without reducing the field strength since the loop's cross-section remains unaffected. If the sequence is repeated n times, the magnetic field strength (and flux) is then amplified by a factor

$$\frac{B^n}{B_0} \propto 2^n = \exp(n \ln 2) \,. \tag{2.21}$$

With n playing the role of a (discrete) time-like variable, Eq. (2.21) indicates an exponential growth of the magnetic field, with a growth rate $\sigma = \ln 2$. Rejoice! This is our first dynamo!

A concept central to the STF dynamo—and other dynamos to be encountered later—is that of *constructive folding*. Note how essential the twisting step is to the STF dynamo: without it (or with an even number of twists), the magnetic field in each half of the folded rope would end up pointing in opposite direction, and would then add up to zero net flux, a case of *destructive folding*. We'll have more to say on the STF dynamo later on; for now we turn to amplification by fluid motions that shear rather than stretch.

2.2.3 Hydrodynamical Shearing and Field Amplification

Magnetic field amplification by stretching, as illustrated on Fig. 2.3, evidently requires (1) a stagnation point in the flow, and (2) a rather specific positioning and orientation of the flux tube with respect to this stagnation point. We will encounter later more realistic flows that do achieve field amplification through stretching in the vicinity of stagnation points, but there is a different type of fluid motion that can produce a more robust form of magnetic field amplification: *shearing*.

The idea is illustrated in cartoon form on Fig. 2.6. We start with a magnetic flux tube, as before, but this time the flow is everywhere perpendicular to the axis of the tube, and its magnitude varies with height along the length of the tube, e.g., $\boldsymbol{u} = u_x(z)\hat{\boldsymbol{e}}_x$, so that $\partial u_x/\partial z \neq 0$; this is called a *planar sheared flow*. In the ideal MHD limit, every small section of the tube is displaced sideways at a rate equal to the flow speed. After a while, the tube will no longer be straight, and for the type of shearing motion illustrated on Fig. 2.6, its length will have increased. By the same logic as before, the field strength within the tube must have increased proportionally to the increase in length of the tube. Note also that here the action of the shear leaves the z-component of the magnetic field unaffected, but produces an x-component where there was initially none.

The beauty of this mechanism is that it does not require stagnation points, and will in general operate for any sheared flow. The latter turn out to be rather common in astrophysical objects, and we now turn to one particularly important example.

2.2.4 Toroidal Field Production by Differential Rotation

A situation of great (astro)physical interest is the induction of a toroidal magnetic field via the shearing of a pre-existing poloidal magnetic field threading a differentially rotating sphere of electrically conducting fluid. Working now in spherical polar coordinates (r, θ, ϕ) and assuming overall axisymmetry (i.e., the poloidal field and differential rotation share the same symmetry axis), the flow velocity can be written as:

$$\boldsymbol{u}(r, \theta) = \varpi \Omega(r, \theta)\hat{\boldsymbol{e}}_\phi , \qquad (2.22)$$

2.2 Magnetic Field Amplification by Stretching and Shearing

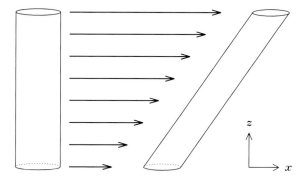

Fig. 2.6 Stretching of a flux tube by a shearing motion directed perpendicularly to the tube's axis. Unlike on Fig. 2.3, here the tube does not remain straight, but it's length is still increasing as a consequence of the tube's deformation; consequently, in the ideal MHD limit the magnetic field threading the tube will be amplified by a factor given by the ratio of its final to initial length, as per Eq. (2.19).

where again $\varpi = r\sin\theta$, and the angular velocity $\Omega(r,\theta)$ is assumed steady ($\partial/\partial t = 0$), corresponding to the kinematic regime introduced earlier. Once again, we take advantage of the poloidal/toroidal separation for axisymmetric magnetic fields (introduced in Sect. 2.1). For the (divergenceless) azimuthal flow $\boldsymbol{u}$ given by the above expression, $\nabla \times (\boldsymbol{u} \times \boldsymbol{B}) = (\boldsymbol{B}\cdot\nabla)\boldsymbol{u} - (\boldsymbol{u}\cdot\nabla)\boldsymbol{B}$, which for an axisymmetric magnetic field only has a non-zero contribution in the ϕ-direction since $\boldsymbol{u}$ itself is azimuthally-directed.[2]

If moreover one neglects magnetic dissipation, the induction equation now separates into:

$$\frac{\partial A}{\partial t} = 0, \qquad (2.23)$$

$$\frac{\partial B}{\partial t} = \varpi[\nabla \times (A\hat{\boldsymbol{e}}_\phi)] \cdot \nabla\Omega. \qquad (2.24)$$

Equation (2.23) states that the poloidal component remains constant in time, so that Eq. (2.24) integrates immediately to

$$B(r,\theta,t) = B(r,\theta,0) + \left(\varpi[\nabla \times (A\hat{\boldsymbol{e}}_\phi)] \cdot \nabla\Omega\right)t. \qquad (2.25)$$

Anywhere in the domain, the toroidal component of the magnetic field grows linearly in time, at a rate proportional to the net local shear and local poloidal field strength. A toroidal magnetic component is being generated by shearing the

---

[2] In trying to demonstrate this, keep in mind that $\boldsymbol{B}\cdot\nabla$ and $\boldsymbol{u}\cdot\nabla$ are differential operators acting on *vector* quantities; it may come as a surprise to realize that $(\boldsymbol{B}\cdot\nabla)\boldsymbol{u}$ has non-vanishing r- and θ-components, even though $\boldsymbol{u}$ is azimuthally-directed! See Appendix A for the component form of these operators in spherical polar coordinates.

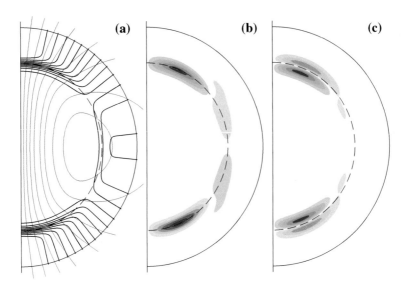

Fig. 2.7 Shearing of a poloidal field into a toroidal component by a solar-like differential rotation profile. Part **a** shows isocontours of the rotation rate $\Omega(r,\theta)/2\pi$ (*solid lines*, contour spacing 10 nHz). The *red lines* are fieldlines for the $n = 1$ dipolar diffusive eigenmode with core-to-envelope diffusivity contrast $\Delta \eta = 10^{-2}$. The *dashed line* is the core–envelope interface at $r/R = 0.7$. Part **b** shows isocontours of the toroidal field, with *yellow-red* (*green-blue*) corresponding to positive (negative) B, after 10 yr of shearing. The maximum toroidal field strength is about 0.2 T, and contour spacing is 0.02 T. Part **c** shows logarithmically spaced isocontours of the ϕ-component of the Lorentz force associated with the poloidal/toroidal fields of panels (**a**) and (**b**).

initially purely poloidal fieldlines in the ϕ-direction, and the magnitude of the poloidal magnetic component remains unaffected, as per Eq. (2.23). Note also that for such an axisymmetric configuration, the only possible steady-state ($\partial/\partial t = 0$) solutions must have

$$[\nabla \times (A\hat{\mathbf{e}}_\phi)] \cdot \nabla \Omega = 0 , \qquad (2.26)$$

i.e., the angular velocity must be constant on any given poloidal flux surface. This result is known as *Ferraro's theorem*.

Evidently, computing B via Eq. (2.25) requires a knowledge of the solar internal (differential) rotation profile $\Omega(r, \theta)$. Consider the following parametrization:

$$\Omega(r,\theta) = \Omega_C + \frac{\Omega_S(\theta) - \Omega_C}{2}\left[1 + \mathrm{erf}\left(\frac{r - r_C}{w}\right)\right] , \qquad (2.27)$$

where

$$\Omega_S(\theta) = \Omega_{Eq}(1 - a_2 \cos^2\theta - a_4 \cos^4\theta) \qquad (2.28)$$

is the surface latitudinal differential rotation. We will make repeated use of this parametrization in this and following chapters, so let's look into it in some detail.

2.2 Magnetic Field Amplification by Stretching and Shearing

Figure 2.7a shows isocontours of angular velocity (in black) generated by the above parameterization with parameter values $\Omega_C/2\pi = 432.8$ nHz, $\Omega_{Eq}/2\pi = 460.7$ nHz, $a_2 = 0.1264$, $a_4 = 0.1591$, $r_c = 0.713R$, and $w = 0.05R$, as obtained by a best-fit to helioseismic frequency splittings. This properly reproduces the primary features of full helioseismic inversions, namely:

1. A convective envelope ($r \gtrsim r_c$) where the shear is purely latitudinal, with the equatorial region rotating faster than the poles;
2. A core ($r \lesssim r_c$) that rotates rigidly, at a rate equal to that of the surface mid-latitudes;
3. A smooth matching of the core and envelope rotation profiles occurring across a thin spherical layer coinciding with the core–envelope interface ($r = r_c$), known as the *tachocline*.

It should be emphasized already at this juncture that such a solar-like differential rotation profile is quite complex, in that it is characterized by *three* partially overlapping shear regions: a strong positive radial shear (i.e., $\partial\Omega/\partial r > 0$) in the equatorial regions of the tachocline, an even stronger negative radial shear ($\partial\Omega/\partial r < 0$) in its polar regions, and a significant positive latitudinal shear ($\partial\Omega/\partial\theta > 0$) throughout the convective envelope and extending partway into the tachocline. For a tachocline of half-thickness $w/R = 0.05$, the mid-latitude latitudinal shear at $r/R = 0.7$ is comparable in magnitude to the equatorial radial shear; as we will see in the next chapter, its potential contribution to dynamo action should not be casually dismissed.

Figure 2.7b shows the distribution of toroidal magnetic field resulting from the shearing of the slowest decaying, $n = 1$ dipole-like diffusive eigenmode of Sect. 2.1 of strength 10^{-4} T at $r/R = 0.7$, using the diffusivity profile given by Eq. (2.16) with diffusivity contrast $\Delta\eta = 10^{-2}$ (part a, red lines). This is nothing more than Eq. (2.25) evaluated for $t = 10$ yr, with $B(r, \theta, 0) = 0$. Not surprisingly, the toroidal field is concentrated in the regions of large radial shear, at the core–envelope interface (dashed line). Note how the toroidal field distribution is *antisymmetric* about the equatorial plane, precisely what one would expect from the inductive action of a shear flow that is equatorially symmetric on a poloidal magnetic field that is itself antisymmetric about the equator.

Knowing the distributions of toroidal and poloidal fields on Fig. 2.7 allows us to flirt a bit with dynamics, by computing the Lorentz force. For the axisymmetric magnetic field considered here, the ϕ-component of Eq. (1.75) reduces to:

$$[\boldsymbol{F}]_\phi = \frac{1}{\mu_0\varpi}\boldsymbol{B}_p \cdot \nabla(\varpi B) , \qquad (2.29)$$

with $\boldsymbol{B}_p = \nabla \times (A\hat{\boldsymbol{e}}_\phi)$ the poloidal field. The resulting spatial distribution of $[\boldsymbol{F}]_\phi$ is plotted on Fig. 2.7c. Examine this carefully to convince yourself that the Lorentz force is such as to *oppose* the driving shear. This is an important and totally general property of interacting flows and magnetic fields: the Lorentz force tends to resist the hydrodynamical shearing responsible here for field induction. The ultimate fate of the system depends on whether the Lorentz force becomes dynamically significant before the growth of the toroidal field is mitigated by resistive dissipation: this is likely the case in solar/stellar interiors.

Clearly, the growing magnetic energy of the toroidal field is supplied by the kinetic energy of the rotational shearing motion (this is hidden in the second term on the RHS of Eq. (1.87)). In the solar case, this is an attractive field amplification mechanism, because the available supply of rotational kinetic energy is immense.[3] But don't make the mistake of thinking that this is a dynamo! In obtaining Eq. (2.25) we have completely neglected magnetic dissipation, and remember, the dynamos we are seeking are flows that can amplify and sustain a magnetic field *against* Ohmic dissipation. In fact, neither flux tube stretching (Fig. 2.3) or shearing (Fig. 2.6) is a dynamo either, for the same reason.[4] Nonetheless, shearing of a poloidal field by differential rotation will turn out to be a central component of *all* solar/stellar dynamo models constructed in later chapters. It is also believed to be an important ingredient of magnetic amplification in stellar accretion disks, and even in galactic disks.

2.3 Magnetic Field Evolution in a Cellular Flow

Having examined separately the resistive decay and hydrodynamical induction of magnetic field, we now turn to a situation where both processes operate simultaneously.

2.3.1 A Cellular Flow Solution

Working now in Cartesian geometry, we consider the action of a steady, incompressible ($\nabla \cdot \boldsymbol{u} = 0$) two-dimensional flow

$$\boldsymbol{u}(x, y) = u_x(x, y)\hat{\boldsymbol{e}}_x + u_y(x, y)\hat{\boldsymbol{e}}_y \tag{2.30}$$

on a two-dimensional magnetic field

$$\boldsymbol{B}(x, y, t) = B_x(x, y, t)\hat{\boldsymbol{e}}_x + B_y(x, y, t)\hat{\boldsymbol{e}}_y \,. \tag{2.31}$$

Note that neither the flow nor the magnetic field have a z-component, and that their x and y-components are both independent of the z-coordinate. The flow is said to be *planar* because $u_z = 0$, and has an ignorable coordinate (i.e., translational

---

[3] This may no longer be the case, however, if dynamo action takes place in a thin layer below the base of the convective envelope; see the paper by see the paper by Steiner & Ferris-Mas (2005) in the bibliography of the next chapter, for more on this aspect of the problem.

[4] The STF dynamo (Fig. 2.5) is the lone exception here, in that it remains a dynamo even when magnetic dissipation is brought into the picture; the reasons why are subtle and will be clarified later on.

2.3 Magnetic Field Evolution in a Cellular Flow

symmetry) since $\partial/\partial z \equiv 0$ for all field and flow components. Such a magnetic field can be represented by the vector potential

$$A = A(x, y, t)\hat{e}_z ,\qquad(2.32)$$

where, as usual, $B = \nabla \times A$. Under this representation, lines of constant A in the $[x, y]$ plane coincide with magnetic fieldlines. The only non-trivial component of the induction equation (1.97) is its z-components, which takes the form

$$\frac{\partial A}{\partial t} + u \cdot \nabla A = \eta \nabla^2 A .\qquad(2.33)$$

This is a linear advection-diffusion equation, describing the transport of a passive scalar quantity A by a flow u, and subject to diffusion, the magnitude of which being measured by η. In view of the symmetry and planar nature of the flow, it is convenient to write the 2D flow field in terms of a stream function $\Psi(x, y)$:

$$u(x, y) = u_0 \left(\frac{\partial \Psi}{\partial y}\hat{e}_x - \frac{\partial \Psi}{\partial x}\hat{e}_y \right) .\qquad(2.34)$$

It is easily verified that any flow so defined will identically satisfy the condition $\nabla \cdot u = 0$. As with Eq. (2.32), a given numerical value of Ψ uniquely labels one streamline of the flow. Consider now the stream function

$$\Psi(x, y) = \frac{L}{4\pi} \left(1 - \cos\left(\frac{2\pi x}{L}\right) \right) \left(1 - \cos\left(\frac{2\pi y}{L}\right) \right), \quad x, y \in [0, L] .\qquad(2.35)$$

This describes a counterclockwise cellular flow centered on $(x, y) = (L/2, L/2)$ as shown on Fig. 2.8. The maximal velocity amplitude $\max \|u\| = u_0$ is found along the streamline $\Psi = u_0 L/(2\pi)$, plotted as a thicker line on Fig. 2.8. This streamline is well approximated by a circle of radius $L/4$, and its streamwise circulation period turns out to be $1.065\, \pi L/2u_0$, quite close to what one would expect in the case of a perfectly circular streamline. In what follows, this timescale is denoted τ_c and referred to as the *turnover time* of the flow. Note that both the normal and tangential components of the flow vanish on the boundaries $x = 0, L$ and $y = 0, L$. This implies that the domain boundary is itself a streamline ($\Psi = 0$, in fact), and that every streamline interior to the boundary closes upon itself within the spatial domain.

We now investigate the inductive action of this flow by solving a nondimensional version of Eq. (2.33), by expressing all lengths in units of L, and time in units of L/u_0, so that

$$\frac{\partial A}{\partial t} = -\frac{\partial \Psi}{\partial y}\frac{\partial A}{\partial x} + \frac{\partial \Psi}{\partial x}\frac{\partial A}{\partial y} + \frac{1}{R_m}\left(\frac{\partial^2 A}{\partial x^2} + \frac{\partial^2 A}{\partial y^2} \right), \quad x, y \in [0, L] ,\qquad(2.36)$$

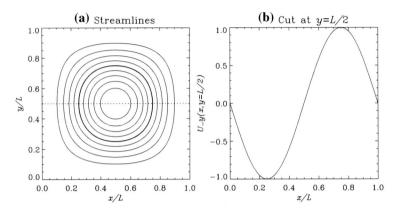

Fig. 2.8 Counterclockwise cellular flow generated by the streamfunction given by Eq. (2.35). Part **a** shows streamlines of the flow, with the thicker streamline corresponding to $\Psi = u_0 L/(2\pi)$, on which the flow attains its maximum speed u_0. Part **b** shows the profile of $u_y(x)$ along a horizontal cut at $y = 1/2$. A "typical" length scale for the flow is then $\sim L$.

where $R_m = u_0 L/\eta$ is the magnetic Reynolds number for this problem, and the corresponding diffusion time is then $\tau_\eta = R_m$ in dimensionless units. Equation (2.36) is solved as an initial-boundary value problem in two spatial dimensions. All calculations described below start at $t = 0$ with an initially uniform, constant magnetic field $\boldsymbol{B} = B_0 \hat{\boldsymbol{e}}_x$, equivalent to:

$$A(x, y, 0) = B_0 y . \tag{2.37}$$

We consider a situation where the magnetic field component normal to the boundaries is held fixed, which amounts to holding the vector potential fixed on the boundary.

Figure 2.9 shows the variation with time of the magnetic energy (Eq. 1.89), for four solutions having $R_m = 10, 10^2, 10^3$ and 10^4. Figure 2.10 shows the evolving shape of the magnetic fieldlines in the $R_m = 10^3$ solution at 9 successive epochs. The solid dots are "floaters", namely Lagrangian markers moving along with the flow. At $t = 0$ all floaters are equidistant and located on the fieldline initially coinciding with the coordinate line $y/L = 0.5$, that (evolving) fieldline being plotted in the same color as the floaters on all panels. Figure 2.10 covers two turnover times.[5]

At first, the magnetic energy increases quadratically in time. This is precisely what one would expect from the shearing action of the flow on the initial B_x-directed magnetic field, which leads to a growth of the B_y-component that is linear in time. However, for $t/\tau_c \gtrsim 2$ the magnetic energy starts to decrease again and eventually $(t/\tau_c \gg 1)$ levels off to a constant value. To understand the origin of this behavior we need to turn to Fig. 2.10 and examine the solutions in some detail.

---

[5] An animation of this solution, and related additional solutions, can be viewed on the course web-page http://obswww.unige.ch/SSAA/sf39/dynamos.

2.3 Magnetic Field Evolution in a Cellular Flow

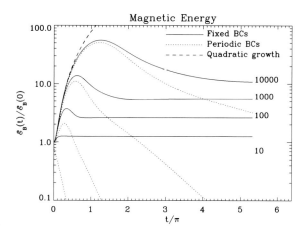

Fig. 2.9 Evolution of the magnetic energy for solutions with different values of R_m. The solutions have been computed over 10 turnover times, at which point they are getting reasonably close to steady-state, at least as far as magnetic energy is concerned. One turnover time corresponds to $t/\pi = 0.532$.

The counterclockwise shearing action of the flow is quite obvious on Fig. 2.10 in the early phases of the evolution, leading to a rather pretty spiral pattern as magnetic fieldlines get wrapped around one another. Note that the distortion of magnetic fieldlines by the flow implies a great deal of *stretching* in the streamwise direction, as well as *folding* in the cross-stream direction. The latter shows up as sharp bends in the fieldlines, while the former is most obvious upon noting that the distance between adjacent floaters increases monotonically in time. In other words, an imaginary flux tube enclosing this fieldline is experiencing the same type of stretching as on Fig. 2.6. It is no accident that the floaters end up in the regions of maximum field amplification on frames 2–5; they are initially positioned on the fieldline coinciding with the line $y = L/2$, everywhere perpendicular to the shearing flow (cf. Figs. 2.6 and 2.8), which pretty much ensures maximal inductive effect, as per Eq. (2.33).

That all floaters remain at first "attached" onto their original fieldline is what one would have expected from the fact that this is a relatively high-R_m solution, so that flux-freezing is effectively enforced. As the evolution proceeds, the magnetic field keeps building up in strength (as indicated by the color scale), but is increasingly confined to spiral "sheets" of decreasing thickness. Coincident with these sheets are strong electrical currents perpendicular to the plane of the page, the current density being given here by

$$J_z(x, y) = -\frac{1}{\mu_0} \nabla^2 A(x, y) \;. \tag{2.38}$$

This current density, integrated over the $[x, y]$ plane, exhibits a time-evolution resembling that of magnetic energy.

By the time we hit one turnover time (corresponding approximately to frame 5 on Fig. 2.10), it seems that we are making progress towards our goal of producing a dynamo; we have a flow field which, upon acting on a preexisting magnetic field, has intensified the strength of that field, at least in some localized regions of the spatial domain. However, beyond $t \sim \tau_c$ the sheets of magnetic fields are gradually

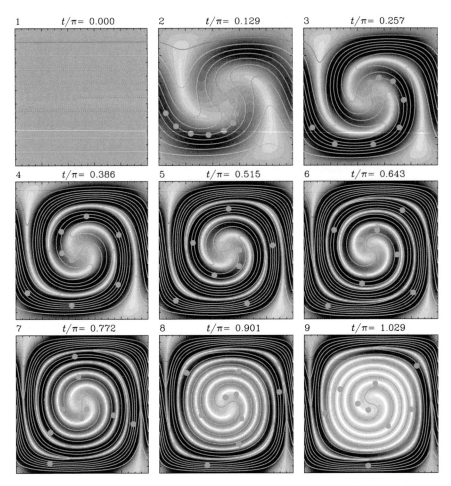

Fig. 2.10 Solution to Eq. (2.36) starting from an initially horizontal magnetic field. The panels show the shape of the magnetic fieldlines at successive times. The color scale encodes the absolute strength of the magnetic field, i.e., $\sqrt{B_x^2 + B_y^2}$. The x- and y-axes are horizontal and vertical, respectively, and span the range $x, y \in [0, L]$. Time t is in units of L/u_0. The *solid dots* are "floaters", i.e., Lagrangian marker passively advected by the flow. The magnetic Reynolds number is $R_m = 10^3$.

disappearing, first near the center of the flow cell (frames 5–7), and later everywhere except close to the domain boundaries (frames 7–9). Notice also how, from frame 5 onward, the floaters are seen to "slip" off their original fieldlines. This means that flux-freezing no longer holds; in other words, diffusion is taking place. Yet, we evidently still have $t \ll \tau_\eta$ ($\equiv R_m = 10^3$ here), which indicates that diffusion should not yet have had enough time to significantly affect the solution. What is going on here?

2.3.2 Flux Expulsion

The solution to this apparent dilemma lies with the realization that we have defined R_m in terms of the global length scale L characterizing the flow. This was a perfectly sensible thing to do on the basis of the flow configuration and initial condition on the magnetic field. However, as the evolution proceeds beyond $\sim \tau_c$ the decreasing thickness of the magnetic field sheets means that the global length scale L is no longer an adequate measure of the "typical" length scale of the magnetic field, which is what is needed to estimate the diffusion time τ_η (see Eq. (1.63)).

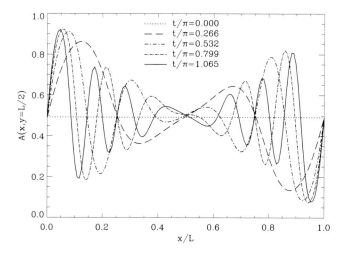

Fig. 2.11 Cuts of a $R_m = 10^4$ solution along the coordinate line $y = 0.5$, at successive times. Note how the "typical" length scale ℓ for the solution decreases with time, from $\ell/L \sim 0.25$ at $t/\pi = 0.266$, down to $\ell/L \sim 0.05$ after two turnover times ($t/\pi = 1.065$).

Figure 2.11 shows a series of cuts of the vector potential A in a $R_m = 10^4$ solution, plotted along the coordinate line $y = L/2$, at equally spaced successive time intervals covering two turnover times. Clearly the inexorable winding of the fieldline leads to a general decrease of the length scale characterizing the evolving solution. In fact, each turnover time adds two new "layers" of alternating magnetic polarity to the spiraling sheet configuration, so that the average length scale ℓ decreases as t^{-1}:

$$\frac{\ell(t)}{L} \propto \frac{L}{u_0 t}, \tag{2.39}$$

which implies in turn that the *local* dissipation time, $\propto \ell^2/\eta$, is decreasing as t^{-2}. On the other hand, examination of Fig. 2.10 soon reveals that the (decreasing) length scale characterizes the thickness of elongated magnetic structures that are themselves more or less *aligned* with the streamlines, so that the turnover time τ_c remains the proper timescale measuring field induction. With τ_c fixed and τ_η inexorably

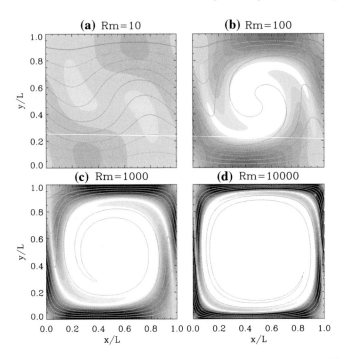

Fig. 2.12 Steady-state solutions to the cellular flow problem, for increasing values of the magnetic Reynolds number R_m. The $R_m = 10^4$ solution is at the resolution limit of the $N_x \times N_y = 128 \times 128$ mesh used to obtain these solutions, as evidenced on part (**d**) by the presence of small scale irregularities where magnetic fieldlines are sharply bent. The color scale encodes the local magnitude of the magnetic field. Note how, in the higher R_m solutions, magnetic flux is expelled from the center of the flow cell. With $\mathcal{E}_B(0)$ denoting the energy of a purely horizontal field with same normal boundary flux distribution, the magnetic energy for these steady states is $\mathcal{E}_B/\mathcal{E}_B(0) = 1.37, 2.80, 5.81$ and 11.75, respectively, for panels (**a**) through (**d**).

decreasing, the solution is bound to reach a point where $\tau_\eta \simeq \tau_c$, *no matter how small dissipation actually is*. To reach that stage just takes longer in the higher R_m solutions, since more winding of the fieldlines is needed. Larger magnetic energy can build up in the transient phase, but the growth of the magnetic field is *always* arrested even in the limit $R_m \to 0$. Equating τ_c ($\sim L/u_0$) to the *local* dissipation time ℓ^2/η, one readily finds that the length scale ℓ at which both process become comparable can be expressed in terms of the *global* R_m as

$$\frac{\ell}{L} = (R_m)^{-1/2}, \qquad R_m = \frac{u_0 L}{\eta}. \tag{2.40}$$

That such a balance between induction and dissipation materializes means that a steady-state can be attained. Figure 2.12 shows four such steady state solutions for increasing values of the (global) magnetic Reynolds number R_m. The higher R_m solutions clearly show *flux expulsion* from the central regions of the domain. This

2.3 Magnetic Field Evolution in a Cellular Flow

is a general feature of steady, high-R_m magnetized flows with closed streamlines: magnetic flux is expelled from the regions of closed streamlines towards the edges of the flow cells, where it ends up concentrated in *boundary layers* which indeed have a thickness of order $R_m^{-1/2}$, as suggested by Eq. (2.40), within which strong z-directed electrical currents flow—and dissipate! It is important to understand how and why this happens.

To first get an intuitive feel for how flux expulsion operates, go back to Fig. 2.10. As the flow wraps the fieldlines around one another, it does so in a manner that folds fieldlines of opposite polarity closer and closer to each other. When two such fieldlines are squeezed closer together than the dissipative length scale (Eq. (2.40)), resistive decay takes over and destroys the field faster than it is being stretched. This is an instance of *destructive folding*, and can only be avoided along the boundaries, where the normal component of the field is held fixed. For flux expulsion to operate, flux-freezing must be effectively enforced on the spatial scale of the flow. Otherwise the field is largely insensitive to the flow, and fieldlines are hardly deformed with respect to their initial configuration (as on panel (a) of Fig. 2.12).

Consider now the implication for the total magnetic flux across the domain; flux conservation requires that the normal flux $B_0 L$ imposed at the right and left boundaries must somehow cross the interior, otherwise Maxwell's equation, $\nabla \cdot \boldsymbol{B} = 0$, would not be satisfied; because of flux expulsion, it can only do so in the thin layers along the bottom and top boundaries. Since the thickness of these layers scales as $R_m^{-1/2}$, it follows that the field strength therein scales as $\sqrt{R_m}$, which in turn implies that the total magnetic energy in the domain also scales as $\sqrt{R_m}$ in the $t \gg \tau_c$ limit.

2.3.3 Digression: The Electromagnetic Skin Depth

You may recall that a sinusoidally oscillating magnetic field imposed at the boundary of a conductor will penetrate the conductor with an amplitude decreasing exponentially away from the boundary and into the conductor, with a length scale called the *electromagnetic skin depth*:

$$\ell = \sqrt{\frac{2\eta}{\omega}}. \tag{2.41}$$

Now, go back to the cellular flow and imagine that you are an observer located in the center of the flow cell, looking at the domain boundaries while rotating with angular frequency $\propto u_0/L$; what you "see" in front of you is an "oscillating" magnetic field, in the sense that it flips sign with "angular frequency" u_0/L. The corresponding electromagnetic skin depth would then be

$$\frac{\ell}{L} = \sqrt{\frac{2\eta}{u_0 L}} \equiv \sqrt{\frac{2}{R_m}}, \tag{2.42}$$

which basically corresponds to the thickness of the boundary layer where significant magnetic field is present in the steady-states shown on Fig. 2.12. How about that for a mind flip...

2.3.4 Timescales for Field Amplification and Decay

Back to our cellular flow. Flux expulsion or not, it is clear from Fig. 2.9 (solid lines) that some level of field amplification has occurred in the high R_m solutions, in the sense that $\mathcal{E}_B(t \to \infty) > \mathcal{E}_B(0)$. But is this a dynamo? The solutions of Fig. 2.12 have strong electric currents in the direction perpendicular to the plane of the paper, concentrated in boundary layers adjacent to the domain boundaries and subjected to resistive dissipation. Have we then reached our goal, namely to amplify and maintain a weak, preexisting magnetic field against Ohmic dissipation?

In a narrow sense yes, but a bit of reflection will show that the boundary conditions are playing a crucial role. The only reason that the magnetic energy does not asymptotically go to zero is that the normal field component is held fixed at the boundaries, which, in the steady-state, implies a non-zero Poynting flux into the domain across the left and right vertical boundaries. The magnetic field is not avoiding resistive decay because of field induction within the domain, but rather because external energy (and magnetic flux) is being pumped in through the boundaries. This is precisely what is embodied in the first term on the RHS of Eq. (1.87).

What if this were not the case? One way to work around the boundary problem is to replace the fixed flux boundary conditions by periodic boundary conditions on B, which in terms of A becomes:

$$A(0, y) = A(L, y), \qquad \frac{\partial A(x, 0)}{\partial y} = \frac{\partial A(x, L)}{\partial y}. \qquad (2.43)$$

There is still a net flux across the horizontal boundaries at $t = 0$, but the boundary flux is now free to decay away along with the solution. Effectively, we now have an infinitely long row of contiguous flow cells, initially threaded by a horizontal magnetic field extending to $\pm\infty$. It is time to reveal that the hitherto unexplained dotted lines on Fig. 2.9 correspond in fact to solutions computed with such boundary conditions, for the same cellular flow and initial condition as before. The magnetic energy now decays to zero, confirming that the boundaries indeed played a crucial role in the sustenance of the magnetic field in our previous solutions. What is noteworthy is the rate at which it does so. In the absence of the flow and with freely decaying boundary flux, the initial field would diffuse away on a timescale $\tau_\eta \sim L^2/\eta$, which is equal to R_m in units of L/u_0. With the flow turned on, the decay proceeds at an accelerated rate because of the inexorable decrease of the typical length scale associated with the evolving solution, which we argued earlier varied as t^{-1}. What then is the typical timescale for this enhanced dissipation? The decay phase of the field (for $t \gg L/u_0$) is approximately described by

2.3 Magnetic Field Evolution in a Cellular Flow

$$\frac{\partial A}{\partial t} = \eta \nabla^2 A . \tag{2.44}$$

An estimate for the dissipation timescale can be obtained once again via dimensional analysis, by replacing ∇^2 by $1/\ell^2$, as in Sect. 2.1 but with the important difference that ℓ is now a function of time:

$$\ell \to \ell(t) = \left(\frac{L}{t}\right)\left(\frac{L}{u_0}\right), \tag{2.45}$$

in view of our previous discussion (cf. Fig. 2.11 and accompanying text). This leads to

$$\frac{\partial A}{\partial t} \simeq -\frac{\eta u_0^2 t^2}{L^4} A , \tag{2.46}$$

where the minus sign is introduced "by hand" in view of the fact that $\nabla^2 A < 0$ in the decay phase. Equation (2.46) integrates to

$$\frac{A(t)}{A_0} = \exp\left[-\frac{\eta u_0^2}{3L^4} t^3\right] = \exp\left[-\frac{1}{3R_m}\left(\frac{u_0^3 t^3}{L^3}\right)\right] . \tag{2.47}$$

This last expression indicates that with t measured in units of L/u_0, the decay time scales as $R_m^{1/3}$. This is indeed a remarkable situation: in the low magnetic diffusivity regime (i.e., high R_m), the flow has in fact *accelerated* the decay of the magnetic field, even though large field intensification can occur in the early, transient phases of the evolution. This is not at all what a dynamo should be doing!

As it turns out, flux expulsion is even trickier than the foregoing discussion may have led you to believe! Flux expulsion destroys the mean magnetic field component directed *perpendicular* to the flow streamlines. It cannot do a thing to a mean component oriented *parallel* to streamlines. For completely general flow patterns and initial conditions, the dissipative phase with timescale $\propto R_m^{1/3}$ actually characterizes the approach to a state where the advected trace quantity—here the vector potential A—becomes constant *along each streamline*, at a value $\bar{A}$ equal to the initial value of A averaged on each of those streamlines. For the cellular flow and initial conditions used above, this average turns out to be $\bar{A} = 0.5$ *for every streamline*, so that the $R_m^{1/3}$ decay phase corresponds to the true decay of the magnetic field to zero amplitude. If $\bar{A}$ varies from one fieldline to the next, however, the $R_m^{1/3}$ phase is followed by a third decay phase, which proceeds on a timescale $\sim R_m$, since induction no longer operates ($\boldsymbol{u} \cdot \nabla A = 0$) and the typical length scale for A is once again L. At any rate, even with a more favorable initial condition we have further delayed field dissipation, but we still don't have a dynamo since dissipation will proceed inexorably, at best on the "long" timescale $R_m \times (L/u_0)$.

2.3.5 Flux Expulsion in Spherical Geometry: Axisymmetrization

You may think that the flux expulsion problem considered in the preceding section has nothing to do with any astronomical objects you are likely to encounter in your astrophysical careers. Wrong!

Consider the evolution of a magnetic field pervading a sphere of electrically conducting fluid, with the solar-like differential rotation profile already encountered previously (Sect. 2.2.4 and Eqs. (2.27)–(2.28)), and with the field having initially the form of a dipole whose axis is inclined by an angle Θ with respect to the rotation axis ($\theta = 0$). Such a magnetic field can be expressed in terms of a vector potential having components:

$$A_r(r, \theta, \phi) = 0 , \tag{2.48}$$

$$A_\theta(r, \theta, \phi) = (R/r)^2 \sin\Theta (\sin\beta \cos\phi - \cos\beta \sin\phi) , \tag{2.49}$$

$$A_\phi(r, \theta, \phi) = (R/r)^2 [\cos\Theta \sin\theta - \sin\Theta \cos\theta(\cos\beta \cos\phi + \sin\beta \sin\phi)] , \tag{2.50}$$

where β is the azimuthal angle locating the projection of the dipole axis on the equatorial plane.

Now, the vector potential for an inclined dipole can be written as the sum of two contributions, the first corresponding to an aligned dipole ($\Theta = 0$), the second to a perpendicular dipole ($\Theta = \pi/2$), their relative magnitude being equal to $\tan\Theta$. Since the governing equation is linear, the solution for an inclined dipole can be broken into two independent solutions for the aligned and perpendicular dipoles. The former is precisely what we investigated already in Sect. 2.2.4, where we concluded there that the shearing of an aligned dipole by an axisymmetric differential rotation would lead to the buildup of a toroidal component, whose magnitude would grow linearly in time at a rate set by the magnitude of the shear.

The solution for a perpendicular dipole is in many way similar to the cellular flow problem of Sect. 2.3. You can see how this may be the case by imagining looking from above onto the equatorial plane of the sphere; the fieldlines contained in that plane will have a curvature and will be contained within a circular boundary, yet topologically the situation is similar to the cellular flow studied in the preceding section: the (sheared) flow in the equatorial plane is made of closed, circular streamlines contained within that plane, so that we can expect flux expulsion to occur. The equivalent of the turnover time here is the differential rotation timescale, namely the time for a point located on the equator to perform a full 2π revolution with respect to the poles:

$$\tau_{\text{DR}} = (\Omega_{\text{Equ}} - \Omega_{\text{Pole}})^{-1} . \tag{2.51}$$

For a freely decaying dipole, the perpendicular component of the initial dipole will then be subjected to flux expulsion, and dissipated away, at a rate far exceeding purely diffusive decay in the high R_m limit, as argued earlier.

2.3 Magnetic Field Evolution in a Cellular Flow

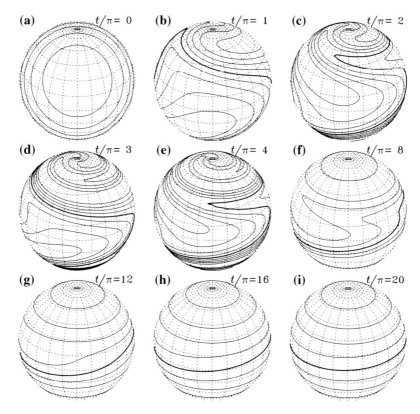

Fig. 2.13 Symmetrization of an inclined dipole in an electrically conducting sphere in a state of solar-like axisymmetric differential rotation. Each panel shows contours of constant B_r at the surface of the sphere, and the solution is matched to a potential in the exterior ($r/R > 1$). The differential rotation is given by Eq. (2.27). Time is given in units of τ_{DR}, in which the turnover period (or differential rotation period) is equal to 2π, and the magnetic Reynolds number is $R_m = 10^3$.

But here is the amusing thing; for an observer looking at the magnetic field at the surface of the sphere, the enhanced decay of the perpendicular component of the dipole will translate into a gradual decrease in the inferred tilt axis of the dipole. Figure 2.13 shows this effect, for the differential rotation profile given by Eq. (2.27) and a magnetic Reynolds number $R_m = 10^3$. Contours of constant B_r are plotted on the surface $r/R = 1$, with the neutral line ($B_r = 0$) plotted as a thicker line. At $t = 0$ the field has the form of a pure dipole tilted by $\pi/3$ with respect to the coordinate axis, and the sphere is oriented so that the observer (you!) is initially looking straight down the magnetic axis of the dipole. Advection by the flow leads to a distortion of the initial field, with the subsequent buildup of small spatial scales in the r- and

θ-directions (only the latter can be seen here).[6] After only two turnover times (last frame), the surface field looks highly axisymmetric.

So, in a differentially rotating fluid system with high R_m, flux expulsion leads to the *symmetrization* of any non-axisymmetric magnetic field component initially present—or contemporaneously generated. The efficiency of the symmetrization process should make us a little cautious in assuming that the large-scale magnetic field of the sun, which one would deem roughly axisymmetric upon consideration of surface things like the sunspot butterfly diagram, is characterized by the same level of axisymmetry in the deep-seated generating layers, where the dynamo is presumed to operate. After all, standing in between is a thick, axisymmetrically differentially rotating convective envelope that must be reckoned with. In fact, observations of coronal density structures in the descending phase of the solar cycle can be interpreted in terms of a large-scale, tilted dipole component, with the tilt angle steadily decreasing over 3–4 years towards solar minimum. Interestingly, the differential rotation timescale for the sun is ~ 6 months. Are we seeing the axisymmetrization process in operation? Maybe. Axisymmetry is certainly a very convenient modelling assumption when working on the large scales of the solar magnetic field, but it may be totally wrong.

Axisymmetrization has also been invoked as an explanation for the almost perfectly axisymmetric magnetic field of the planet Saturn, which stands in stark contrast to the other solar system planetary magnetic fields. Saturn has a very pronounced surface latitudinal differential rotation, characterized by equatorial acceleration, and current structural models suggest that this differential rotation may persist in the molecular Hydrogen envelope, down to the edge of the metallic Hydrogen core ($r/R_S \simeq 0.55$), where dynamo action is presumed to take place. This would be an ideal configuration for axisymmetrization of a non-axisymmetric deep magnetic field, provided the electrical conductivity is high enough at the base of the envelope to ensure good coupling between the magnetic field and the fluid.

2.4 Two Anti-Dynamo Theorems

The cellular flow studied in Sect. 2.3, although it initially looked encouraging (cf. Fig. 2.9), proved not to be a dynamo after all. Is this peculiar to the flow defined by Eqs. (2.34)–(2.35), or is this something more general? Exhaustively testing for dynamo action in all possible kinds of flow geometries is clearly impractical. However, it turns out that one can rule out a priori dynamo action in many classes of flows. These demonstrations are known as *anti-dynamo theorems*.

---

[6] An animation of this solution, as well as a few others for different R_m and/or tilt angle, can be viewed on the course web-page.

2.4 Two Anti-Dynamo Theorems

2.4.1 Zeldovich's Theorem

A powerful anti-dynamo theorem due to Zeldovich (1914–87), has a lot to teach us about our cellular flow results. The theorem rules out dynamo action in incompressible ($\nabla \cdot \boldsymbol{u} = 0$) steady planar flows in cartesian geometry, i.e., flows of the form

$$\boldsymbol{u}_2(x, y, z) = u_x(x, y, z)\hat{\boldsymbol{e}}_x + u_y(x, y, z)\hat{\boldsymbol{e}}_y \tag{2.52}$$

in a bounded volume V at the boundaries (∂V) of which the magnetic field vanishes. Note that no other restrictions are placed on the magnetic field, which can depend on all three spatial coordinate as well as time. Nonetheless, in view of Eq. (2.52) it will prove useful to consider the z-component of the magnetic field $B_z(x, y, z, t)$ separately from the (2D) field component in the $[x, y]$ plane (hereafter denoted $\boldsymbol{B}_2$). It is readily shown that the z-component of the induction equation then reduces to

$$\left(\frac{\partial}{\partial t} + \boldsymbol{u} \cdot \nabla\right) B_z = \eta \nabla^2 B_z \tag{2.53}$$

for spatially constant magnetic diffusivity. Now, the LHS is just a Lagrangian derivative, yielding the time variation of B_z as one moves along with the fluid. Multiplying this equation by B_z and integrating over V yields, after judicious use of Green's first identity (see Appendix A):

$$\frac{1}{2}\int_V \frac{\mathrm{D}B_z^2}{\mathrm{D}t}\mathrm{d}V = \eta \int_{\partial V} B_z(\nabla B_z) \cdot \boldsymbol{n}\, \mathrm{d}S - \eta \int_V (\nabla B_z)^2 \mathrm{d}V \ . \tag{2.54}$$

Now, the first integral on the RHS vanishes since $\boldsymbol{B} = 0$ on ∂V by assumption. The second integral is positive definite, *therefore B_z always decays on the diffusive timescale* (cf. Sect. 2.1).

Consider now the magnetic field $\boldsymbol{B}_2$ in $[x, y]$ planes. The most general such 2D field can be written as the sum of a solenoidal and potential component:

$$\boldsymbol{B}_2(x, y, z, t) = \nabla \times (A\hat{\boldsymbol{e}}_z) + \nabla \Phi \ , \tag{2.55}$$

where the vector potential A and scalar potential Φ both depend on time and on all three spatial coordinates, except for Φ having no z-dependency. The constraint $\nabla \cdot \boldsymbol{B} = 0$ then implies

$$\nabla_2^2 \Phi = -\frac{\partial B_z}{\partial z} \ , \tag{2.56}$$

where $\nabla_2^2 \equiv \partial^2/\partial x^2 + \partial^2/\partial y^2$ is the 2D Laplacian operator in the $[x, y]$ plane. Clearly, once B_z has resistively dissipated, i.e., for times much larger than the global resistive decay time τ_η, Φ is simply a solution of the 2D Laplace equation $\nabla_2^2 \Phi = 0$.

Here comes the sneaky part. We first substitute Eq. (2.55) into the induction (1.59), and then take the curl of the resulting expression; this last manoeuvre will lead to the disappearance of all but one contribution from the $\nabla\Phi$ term, since $\nabla \times \nabla\Phi = 0$ identically. Moreover, since $\boldsymbol{u}_2$ and $A\hat{\boldsymbol{e}}_z$ are here orthogonal by construction, we also have $\boldsymbol{u}_2 \times \nabla \times (A\hat{\boldsymbol{e}}_z) = -(\boldsymbol{u}_2 \cdot \nabla)(A\hat{\boldsymbol{e}}_z)$. Since the time and spatial derivatives commute, a bit of vector algebra allows to write the remaining terms in the form:

$$\nabla \times \nabla \times \left[\frac{\partial (A\hat{\boldsymbol{e}}_z)}{\partial t} + \boldsymbol{u}_2 \cdot \nabla (A\hat{\boldsymbol{e}}_z) - \eta \nabla_2^2 (A\hat{\boldsymbol{e}}_z) - \boldsymbol{u}_2 \times \nabla\Phi \right] = 0 , \qquad (2.57)$$

with $\nabla \cdot (A\hat{\boldsymbol{e}}_z) = 0$ as a choice of gauge. In general, the above expression is only satisfied if the quantity in square brackets itself vanishes, i.e.,

$$\left(\frac{\partial}{\partial t} + \boldsymbol{u}_2 \cdot \nabla \right) A = \eta \nabla_2^2 A + (\boldsymbol{u}_2 \times \nabla\Phi) \cdot \hat{\boldsymbol{e}}_z . \qquad (2.58)$$

This expression is identical to that obtained above for B_z, except for the presence of the source term $\boldsymbol{u}_2 \times \nabla\Phi$. However, we just argued that for $t \gg \tau_\eta$, $\nabla_2^2 \Phi = 0$. In addition, $\boldsymbol{B}$ vanishes on ∂V by assumption, so that the only possible asymptotic interior solutions are of the form $\Phi = $const, which means that the source term vanishes in the limit $t \gg \tau_\eta$. From this point on Eq. (2.58) is indeed identical to Eq. (2.53), for which we already demonstrated the inevitability of resistive decay. Therefore, dynamo action, i.e., maintenance of a magnetic field against resistive dissipation, is impossible in a planar flow for *any* 3D magnetic field.

2.4.2 Cowling's Theorem

Another powerful anti-dynamo theorem, predating in fact Zeldovich's, is due to Cowling (1906–90). This anti-dynamo theorem is particularly important historically, since it rules out dynamo action for 3D but axisymmetric flows and magnetic fields, which happen to be the types of flows and fields one sees in the sun, at least on the larger spatial scales. Rather than going over one of the many formal proofs of Cowling's theorem found in the literature, let's just follow the underlying logic of our proof of Zeldovich's theorem.

Assuming once again that there are no sources of magnetic field exterior to the domain boundaries, we consider the inductive action of a 3D, steady axisymmetric flow on a 3D axisymmetric magnetic field. Working in spherical polar coordinates (r, θ, ϕ), we write:

$$\boldsymbol{u}(r, \theta) = \frac{1}{\varrho} \nabla \times (\Psi(r, \theta)\hat{\boldsymbol{e}}_\phi) + \varpi\Omega(r, \theta)\hat{\boldsymbol{e}}_\phi , \qquad (2.59)$$

$$\boldsymbol{B}(r, \theta, t) = \nabla \times (A(r, \theta, t)\hat{\boldsymbol{e}}_\phi) + B(r, \theta, t)\hat{\boldsymbol{e}}_\phi , \qquad (2.60)$$

2.4 Two Anti-Dynamo Theorems

where $\varpi = r\sin\theta$. Here, the azimuthal vector potential A and stream function Ψ define the poloidal components of the field and flow, and Ω is the angular velocity (units rad s^{-1}). Note that the form of Eq. (2.59) guarantees that $\nabla \cdot (\varrho\boldsymbol{u}) = 0$, describing mass conservation in a steady flow. Separation of the (vector) MHD induction equation into two components for the 2D scalar fields A and B, as done in Sect. 2.1, now leads to:

$$\left(\frac{\partial}{\partial t} + \boldsymbol{u}_p \cdot \nabla\right)(\varpi A) = \varpi\eta\left(\nabla^2 - \frac{1}{\varpi^2}\right) A, \qquad (2.61)$$

$$\left(\frac{\partial}{\partial t} + \boldsymbol{u}_p \cdot \nabla\right)\left(\frac{B}{\varpi}\right) = \frac{\eta}{\varpi}\left(\nabla^2 - \frac{1}{\varpi^2}\right) B + \frac{1}{\varpi^2}\frac{d\eta}{dr}\frac{\partial(\varpi B)}{\partial r}$$
$$- \left(\frac{B}{\varpi}\right)\nabla \cdot \boldsymbol{u}_p + \boldsymbol{B}_p \cdot \nabla\Omega, \qquad (2.62)$$

where $\boldsymbol{B}_p$ and $\boldsymbol{u}_p$ are notational shortcuts for the poloidal field and meridional flow. Notice that the vector potential A evolves in a manner entirely independent of the toroidal field B, the latter being conspicuously absent on the RHS of Eq. (2.61). This is not true of the toroidal field B, which is well aware of the poloidal field's presence via the $\nabla\Omega$ shearing term.

The LHS of these expressions is again a Lagrangian derivative for the quantities in parentheses, and the first terms on each RHS are of course diffusion. The next term on the RHS of Eq. (2.62) vanishes for incompressible flows, and remains negligible for very subsonic compressible flows. The last term on the RHS, however, is a source term, in that it can lead to the growth of B *as long as A does not decay away*. This is the very situation we have considered in Sect. 2.2.4, by holding A fixed as per Eq. (2.23). However, there is no similar source-like term on the RHS of Eq. (2.61), which governs the evolution of A.

This should now start to remind you of Zeldovich's theorem. In fact, Eq. (2.3) is structurally identical to Eq. (2.53), for which we demonstrated the inevitability of resistive decay in the absence of sources exterior to the domain. This means that A will inexorably decay, implying in turn that B will then also decay once A has vanished. Since axisymmetric flows cannot maintain A against Ohmic dissipation, *a 3D axisymmetric flow cannot act as a dynamo for a 3D axisymmetric magnetic field.*[7] Cowling's theorem is not restricted to spherical geometry, and is readily generalized to any situation where both flow and field showing translational symmetry in one and the same spatial coordinate. Such physical systems are said to have an *ignorable coordinate*.

It is worth pausing and reflecting on what these two antidynamo theorems imply for the cellular flow of Sect. 2.3. It was indeed a planar flow ($u_z = 0$), and moreover the magnetic field had an ignorable coordinate ($\partial\boldsymbol{B}/\partial z \equiv 0$)! We thus fell under the

---

[7] A fact often unappreciated is that Cowling's theorem does not rule out the dynamo generation of a *non-axisymmetric* 3D magnetic field by a 3D axisymmetric flow.

purview of both Zeldovich's and Cowling's theorems, so in retrospect our failure to find dynamo action is now understood.

2.5 The Roberts Cell Dynamo

Clearly, the way to evade both theorems is to consider flows and fields that are fully three-dimensional, and lack translational symmetry in either the flow or the magnetic field. We now consider one such flow, and examine some of its dynamo properties.

2.5.1 The Roberts Cell

The Roberts cell is a spatially periodic, incompressible flow defined over a 2D domain $(x, y) \in [0, 2\pi]$ in terms of a stream function

$$\Psi(x, y) = \cos x + \sin y, \qquad (2.63)$$

so that

$$\boldsymbol{u}(x, y) = \frac{\partial \Psi(x, y)}{\partial y}\hat{\boldsymbol{e}}_x - \frac{\partial \Psi(x, y)}{\partial x}\hat{\boldsymbol{e}}_y + \Psi(x, y)\hat{\boldsymbol{e}}_z . \qquad (2.64)$$

Note that the flow velocity is independent of the z-coordinate, even though the flow has a non-zero z-component. Equations (2.63)–(2.64) describe a periodic array of counterrotating flow cells in the $[x, y]$ plane, with a z-component that changes sign from one cell to the next; the total flow is then a series of helices, which have the same kinetic helicity $h = \boldsymbol{u} \cdot \nabla \times \boldsymbol{u}$ in each cell. The Roberts cell flow represents one example of a *Beltrami flows*, i.e., it satisfies the relation $\nabla \times \boldsymbol{u} = \alpha \boldsymbol{u}$, where α is a numerical constant. Such flows are *maximally helical*, in the sense that their vorticity ($\boldsymbol{\omega} \equiv \nabla \times \boldsymbol{u}$) is everywhere parallel to the flow, which maximizes helicity for a given flow speed. Figure 2.14 shows one periodic "unit" of the the Roberts cell flow pattern. Take note already of the presence of stagnation points where the corners of four contiguous flow cells meet.

Let's first pause and consider why one should expect the Roberts cell to evade Cowling's and Zeldovich's theorems. First, note that this is not a planar flow in the sense demanded by Zeldovich's theorem, since we do have three non-vanishing flow components. However, the z-coordinate is ignorable in the sense of Cowling's theorem, since all flow components are independent of z. *If this flow is to evade Cowling's theorem and act as a dynamo, it must act on a magnetic field that is dependent on all three spatial coordinates.*

Consequently, we consider the inductive effects of this flow acting on a fully three dimensional magnetic field $\boldsymbol{B}(x, y, z, t)$. Since the flow speed is independent of z, we can expect solutions of the linear induction equation to be separable in z, i.e.:

2.5 The Roberts Cell Dynamo

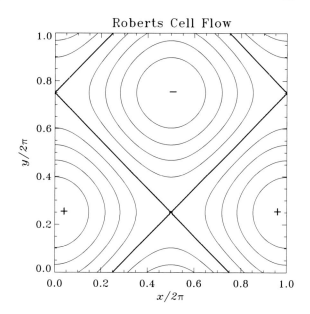

Fig. 2.14 The Roberts cell flow. The flow is periodic in the $[x, y]$ plane, and independent of the z-coordinate (but $u_z \neq 0$!). Flow streamlines are shown projected in the $[x, y]$ plane, and the $+/-$ signs indicate the direction of the z-component of the flow. The *thicker contour* defines the network of *separatrix surfaces* in the flow, corresponding to cell boundaries and intersecting at stagnation points. The $u_z(x, y)$ isocontours coincide with the projected streamlines.

$$\boldsymbol{B}(x, y, z, t) = \boldsymbol{b}(x, y, t)e^{ikz} , \qquad (2.65)$$

where k is a (specified) wave vector in the z-direction, and the 2D magnetic amplitude $\boldsymbol{b}$ is now a complex quantity. We are still dealing with a fully 3D magnetic field, but the problem has been effectively reduced to two spatial dimensions (x, y), which represents a great computational advantage.

2.5.2 Dynamo Solutions

From the dynamo point of view, the idea is again to look for solutions of the induction equations where the magnetic energy does not fall to zero as $t \to \infty$. In practice this means specifying k, as well as some weak field as an initial condition, and solve the 2D linear initial value problem for $\boldsymbol{b}(x, y, t)$ resulting from the substitution of Eq. (2.65) into the induction equation:

$$\frac{\partial \boldsymbol{b}}{\partial t} = (\boldsymbol{b} \cdot \nabla_{xy})\boldsymbol{u} - (\boldsymbol{u} \cdot \nabla_{xy})\boldsymbol{b} - iku_z\boldsymbol{b} + \mathrm{R}_{\mathrm{m}}^{-1}(\nabla_{xy}^2\boldsymbol{b} - k^2\boldsymbol{b}) , \qquad (2.66)$$

subjected to periodic boundary conditions on $\boldsymbol{b}$, in order to avoid the potentially misleading role of fixed-flux boundary conditions in driving dynamo action, as encountered in Sect. 2.3. Here ∇_{xy} and ∇_{xy}^2 are the 2D gradient and Laplacian operators in the $[x, y]$ plane. As before we use as a time unit the turnover time τ_{c}, which

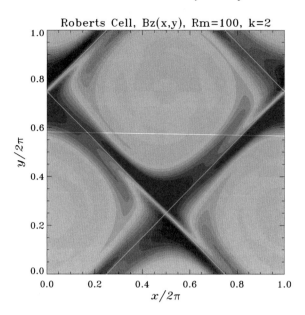

Fig. 2.15 Isocontours for the z-component of the magnetic field in the $[x, y]$ plane, for Roberts cell dynamo solutions with $R_m = 100$ and $k = 2$, in the asymptotic regime $t \gg \tau_c$. The color scale codes the real part of the z-component of $\boldsymbol{b}(x, y, t)$ (*gray-to-blue* is negative, *gray-to-red* positive). The *green straight lines* indicate the separatrix surfaces of the flow (see Fig. 2.14). Note the flux expulsion from the cell centers, and the concentration of the magnetic flux in thin sheets pressed against the separatrices. In the $t \gg \tau_c$ regime, the field grows exponentially but the shape of the planform is otherwise steady. Compare this with Fig. 2.12b.

is of order 2π here. All solutions described below were again obtained numerically, starting from a weak, horizontal magnetic field as the initial condition.

The time evolution can be divided into three more or less distinct phases, the first two being similar to the case of the 2D cellular flow considered in Sect. 2.3: (1) quadratic growth of the magnetic energy for $t \lesssim \tau_c$; (2) flux expulsion for the subsequent few τ_c. However, and unlike the case considered in Sect. 2.3, for some values of k the third phase is one of *exponential growth* in the magnetic field (and energy).

Figure 2.15 shows a typical Roberts cell dynamo solution, for $R_m = 10^2$ and $k = 2$. What is plotted is the real part of the z-component of $\boldsymbol{b}(x, y, t)$, at time $t \gg \tau_c$. The thick green lines are the separatrices of the flow. One immediately recognizes the workings of flux expulsion, in that very little magnetic flux is present near the center of the flow cells. Instead the field is concentrated in boundary-layer-like thin sheets parallel to the separatrix surfaces. Given our extensive discussion of flux expulsion in Sect. 2.3, it should come as no surprise that the thickness of those sheets scales as $R_m^{-1/2}$. For $t \gg \tau_c$, the field grows exponentially, but the shape of the "planform" remains fixed. In other words, even though we solved the induction

2.5 The Roberts Cell Dynamo

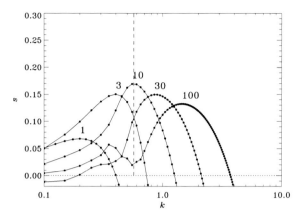

Fig. 2.16 Growth rates of the magnetic energy in the Roberts cell, for sequences of solutions with increasing k and various values of R_m, as labeled near the maxima of the various curves. Growth typically occurs for a restricted range in k, and peaks at a value k_{max} that increases slowly with increasing R_m. Note however how the corresponding maximum growth rate decreases with increasing R_m. The small "dip" left of the main peaks for the high-R_m solutions is a real feature, although here it is not very well resolved in k.

equation as an initial value problem, the solution can be thought of as an eigensolution of the form $\boldsymbol{B}(x, y, z, t) = \boldsymbol{b}(x, y)e^{ikz+st}$, with $\mathrm{Re}(s) > 0$ and $\mathrm{Im}(s) = 0$.

In terms of the magnetic energy evolution, the growth rate s of $\boldsymbol{b}(x, y, t)$ is readily obtained by a linear least-squares fit to the $\log \mathcal{E}_B$ versus t curves in the $t \gg \tau_c$ regime, or more formally defined as

$$s = \lim_{t \to \infty} \left[\frac{1}{2t} \log(\mathcal{E}_B) \right]. \tag{2.67}$$

It turns out that the Roberts cell flows yields dynamo action (i.e., $s > 0$) over wide ranges of wave numbers k and magnetic Reynolds number R_m. Figure 2.16 shows the variations in growth rates with k, for various values of R_m. The curves peak at a growth rate value k_{max} that gradually shifts to higher k as R_m increases. The largest growth rate is $k_{max} \simeq 0.17$, and occurs at $R_m \simeq 10$. It can be shown (see bibliography) that in the high R_m regimes the following scalings hold:

$$k_{max} \propto R_m^{1/2}, \quad R_m \gg 1, \tag{2.68}$$

$$s(k_{max}) \propto \frac{\log(\log R_m)}{\log R_m}, \quad R_m \gg 1. \tag{2.69}$$

To understand the origin of these peculiar scaling relations, we need to take a closer look at the mechanism through which the magnetic field is amplified by the Roberts cell.

2.5.3 Exponential Stretching and Stagnation Points

Even cursory examination of Fig. 2.15 suggests that magnetic field amplification in the Roberts cell is somehow associated with the network of separatrices and stagnation points. It will prove convenient in the foregoing analysis and discussion to first introduce new coordinates:

$$x' = x - y, \qquad y' = x + y + \frac{3\pi}{2}, \tag{2.70}$$

corresponding to a $3\pi/2$ translation in the y-direction, followed by 45° rotation about the origin in the $[x, y]$ plane. The separatrices are now parallel to the coordinate lines $x' = n\pi$, $y' = n\pi$ ($n = 0, 1, \ldots$), and the stream function has become

$$\Psi(x', y') = 2\sin(x')\sin(y'). \tag{2.71}$$

Close to the stagnation points, a good approximation to Eq. (2.71) is

$$\Psi(x', y') \simeq 2x'y', \qquad x', y' \ll 1, \tag{2.72}$$

which, if anything else, should now clarify why this is called a hyperbolic stagnation point... Consider now a fluid element flowing in the vicinity of this stagnation point. From a Lagrangian point of view its equations of motion are:

$$\frac{\partial x'}{\partial t} = u_{x'} = 2x', \tag{2.73}$$

$$\frac{\partial y'}{\partial t} = u_{y'} = -2y', \tag{2.74}$$

which immediately integrates to

$$x'(t) = x'_0 e^{2t}, \qquad y'(t) = y'_0 e^{-2t}, \tag{2.75}$$

where (x'_0, y'_0) is the location of the fluid element at $t = 0$. Evidently, the fluid element experiences *exponential stretching* in the x'-direction, and corresponding contraction in the y'-direction (since $\nabla \cdot \boldsymbol{u} = 0$!). Now, recall that in ideal MHD ($R_m = \infty$) a magnetic fieldline obeys an equation identical to that of a line element, and that stretching leads to field amplification as per the mass conservation constraint (Sect. 2.2.1). Clearly, stagnation points have quite a bit of potential, when it comes to amplifying exponentially a pre-existing magnetic field... provided diffusion and destructive folding can be held at bay. Let's look into how this is achieved in the Roberts Cell.

2.5.4 Mechanism of Field Amplification in the Roberts Cell

We stick to the rotated Roberts cell used above, restrict ourselves to the $R_m \gg 1$ regime, and pick up the field evolution after flux expulsion is completed and the magnetic field is concentrated in thin boundary layers (thickness $\propto R_m^{-1/2}$) pressed against the separatrices (as on Fig. 2.15).

Consider a x'-directed magnetic fieldline crossing a vertical separatrix, as shown on Fig. 2.17a (gray line labeled "a"). The y' component of the flow is positive on either side of the separatrix, and peaks on the separatrix. Consequently, the fieldline experiences stretching in the y'-direction ($a \to b \to c \to d$ on Fig. 2.17a). However, the induced y' component of the magnetic field changes sign across the separatrix, so that we seem to be heading towards our dreaded destructive folding. This is where the crucial role of the vertical (z) dimension becomes apparent. Figure 2.17b is a view of the same configuration in the $[x', z]$ plane, looking down onto the y' axis on part A. At $t = 0$ the fieldlines have no component in the z-direction, but in view of the assumed e^{ikz} spatial dependency the x' component changes sign every half-wavelength k/π. Consider now the inductive action of the z-component of the velocity, which changes sign across the separatrix. After some time interval of order $k/(\pi u_z)$ the configuration

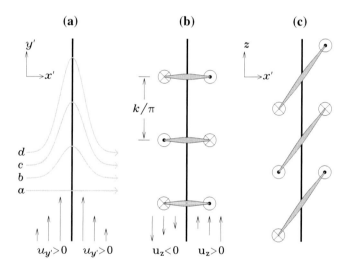

Fig. 2.17 Mechanism of magnetic field amplification in the Roberts cell flow. The diagram is plotted in terms of the rotated $[x', y']$ Roberts cell. The *thick vertical line* is a separatrix surface, and the *gray lines* are magnetic fieldlines. Part **a** is a view in the horizontal plane $[x', y']$, and shows the production of a y'-directed magnetic component from an initially x'-directed magnetic field (line labeled "a"). Parts **b** and **c** are views in the $[x', z]$ plane looking down along the y' axis, and illustrate the phase shift in the z-direction of the y' magnetic component caused by the z-component of the velocity. The symbol $\odot$ ($\otimes$) indicates a magnetic field coming out (into) the plane of the page. Note on part (**c**) how footpoints of identical polarity are brought in close proximity, thus avoiding the destructive folding that would have otherwise characterized the situation depicted on part (**b**) in the $u_z = 0$ 2D case.

of Fig. 2.17b will have evolved to that shown on part C. Observe what has happened: the fieldlines have been sheared in such a way that y'-components of the magnetic field of like signs have been brought in close proximity. Contrast this to the situation on part B, where magnetic footpoints in closest proximity have oppositely directed y'-components.

The end result of this process is that a y'-directed magnetic field is produced by shearing of the initial x'-directed field, with a phase shift in the z-direction such that destructive folding is avoided. Clearly, this requires *both* a z-component of velocity, *and* a z-dependency in the magnetic field. Either alone won't do the trick.

Now, the same reasoning evidently applies to a y'-directed magnetic fieldline crossing a horizontal separatrix: a x'-directed magnetic field will be induced. That magnetic field will be swept along the horizontal separatrix, get further amplified by exponential stretching as it zooms by the stagnation point, and continue along the vertical separatrix, where it can now serve as a seed field for the production of a y'-directed field. The dynamo "loop" is closed, at any time the rate of field production is proportional to the local field strength, and exponential growth of the field follows. The process works best if the half wavelength k/π is of order of the boundary layer thickness, which in fact is what leads to the scaling law given by Eq. (2.68). The scaling for the growth rate (Eq. 2.69), in turn, is related to the time spent by a fluid element in the vicinity of the stagnation point.

2.5.5 Fast Versus Slow Dynamos

One worrisome aspect of the Roberts cell dynamo is the general decrease of the growth rates with increasing R_m (see Fig. 2.16); worrisome, because the $R_m \to \infty$ limit is the one relevant to most astrophysically interesting circumstances. A dynamo exhibiting this property is called a *slow dynamo*, in contrast to a *fast dynamo*, which (by definition) retains a finite growth rate as $R_m \to \infty$, the formal requirement being that

$$\lim_{R_m \to \infty} s(k_{\max}) > 0 . \tag{2.76}$$

In view of Eq. (2.69), the Roberts cell is thus formally a slow dynamo. However the RHS of Eq. (2.69) is such a slowly decreasing function of R_m that the Roberts cell is arguably the closest thing it could be to a fast dynamo... without formally being one.

The distinction hinges on the profound differences between the strict mathematical case of $R_m = \infty$ (ideal MHD), and the more physically relevant limit $R_m \to \infty$. From the physical point of view, the distinction is a crucial one. One example will suffice. Recall that in the absence of dissipation magnetic helicity is a conserved quantity in any evolving magnetized fluid:

2.5 The Roberts Cell Dynamo

$$\frac{d\mathcal{H}_B}{dt} = \frac{d}{dt}\int_V \boldsymbol{A} \cdot \boldsymbol{B}\, dV = 0, \tag{2.77}$$

where $\boldsymbol{B} = \nabla \times \boldsymbol{A}$. Dynamo action, in the sense of amplifying a weak initial field, is then clearly impossible except for the subset of initial fields having $\mathcal{H}_B = 0$. This is a very stringent constraint on dynamo action! Go back now to the Roberts cell dynamo in the high-R_m regime. We saw that magnetic structures build up on a horizontal length scale $\propto R_m^{-1/2}$, and that the vertical wavelength of the fastest growing mode also decreases as $R_m^{-1/2}$. *The inexorable shrinking of the length scales ensures that dissipation always continue to operate even in the $R_m \to \infty$ limit.* This is why the Roberts cell dynamo can evade the constraint of helicity conservation. This is also why it is a slow dynamo. On the other hand, the Vainshtein & Zeldovich Stretch–Twist–Fold dynamo of Sect. 2.2, with its growth rate $\sigma = \ln 2$, is a fast dynamo since nothing prevents it from operating in the $R_m \to \infty$ limit.

But is this really the case? In the flows we have considered up to now, the existence of dynamo action hinges on stretching winning over destructive folding; in the 2D cellular flow of Sect. 2.3, destructive folding won over stretching everywhere away from boundaries. In the Roberts cell, destructive folding is avoided only for vertical wave numbers such that magnetic fields of like signs are brought together, minimizing dissipation. The STF dynamo actually combines stretching and constructive folding, such that folding *reinforces* stretching. The fact that destructive folding is avoided entirely is why the growth rate does not depend on R_m.

Well, upon further consideration it turns out that magnetic diffusivity must play a role in the STF rope dynamo after all. Diffusion comes in at two levels; the first and most obvious one is at the "crossings" formed by the STF sequence. The second and less obvious arises from the fact that as one applies the STF operation n times, the resulting "flux rope" is in fact made up of n closely packed flux ropes, each of cross-section $\propto 2^{-n}$ times smaller than the original circular flux rope, so that the total cross-section looks more like a handful of spaghettis that it does a single monolithic flux rope of strength $\propto 2^n$. If one waits long enough, the magnetic length scale perpendicular to the loop axis shrinks to zero, so that even in the $R_m \to \infty$ limit dissipation is bound to come into to play.

2.6 The CP Flow and Fast Dynamo Action

It turns out that a simple modification of the Roberts cell flow can turn it into a true fast dynamo. The so-called CP flow (for "Circularly Polarized") is nothing more that the original Roberts cell flow, with a forced time-dependence. It is once again a spatially periodic, incompressible flow, defined in cartesian coordinate over a 2D domain $(x, y) \in [0, 2\pi]$:

$$u_x(x, y, t) = A\cos(y + \varepsilon \sin \omega t), \tag{2.78}$$
$$u_y(x, y, t) = C\sin(x + \varepsilon \cos \omega t), \tag{2.79}$$
$$u_z(x, y, t) = A\sin(y + \varepsilon \sin \omega t) + C\cos(x + \varepsilon \cos \omega t). \tag{2.80}$$

Although the CP flow is not expressed here in terms of a stream function, this is the same as the Roberts Cell flow, except that now the counter-rotating flow cells are "precessing" in unison in the $[x, y]$ plane, along circular paths of radius ε, undergoing a full revolution in a time interval $2\pi/\omega$. Here and in what follows we set $\omega = 1$, $\varepsilon = 1$, $A = C = \sqrt{3/2}$, without any loss of generality.

2.6.1 Dynamo Solutions

The CP flow has the same spatial symmetry properties as the Roberts cell, and in particular is invariant in the z-direction. Consequently we again need to seek magnetic solutions with a z-dependency to evade Cowling's theorem. The magnetic field is again separable in z (Eq. 2.65), which leads to the 2D form of the induction equation already encountered with the Roberts cell (Eq. 2.66), subjected to periodic boundary conditions on $\boldsymbol{b}(x, y, t)$. As before, the idea is to pick a value for the vertical wavenumber k, and monitor dynamo action by tracking the growth (or decay) of the magnetic energy via Eq. (2.67).

Computing solutions for varying k soon reveals that dynamo action (i.e., positive growth rates $s(k, \mathrm{R_m})$) occurs in a finite range of vertical wavenumber k, with exponential growth setting in after a time of order of the turnover time. Figure 2.18 shows a snapshot of the vertical magnetic field $b_z(x, y, t)$ in this phase of exponential growth, for a $\mathrm{R_m} = 2000$ solution with $k = 0.57$, which here yields the largest growth rate. The solution is fully time-dependent, and its behavior is best appreciated by viewing it as an animation.[8]

The solution is characterized by multiple sheets of intense magnetic field, of thickness once again $\propto \mathrm{R_m}^{-1/2}$. The magnetic field exhibits *spatial intermittency*, in the sense that if one were to randomly choose a location somewhere in the $[x, y]$ plane, chances are good that only a weakish magnetic field would be found. In high-$\mathrm{R_m}$ solutions, strong fields are concentrated in small regions of the domain; in other words, their *filling factor* is small. This can be quantified by computing the *probability density function* (hereafter PDF) of the magnetic field strength, $f(|B_z|)$. This involves measuring B_z at every (x, y) mesh point in the solution domain, and simply counting how many mesh points have $|B_z|$ between values B and $B + \mathrm{d}B$. The result of such a procedure is shown in histogram form on Fig. 2.19. The PDF shows a power-law tail at high field strengths,

$$f(|B_z|) \propto |B_z|^{-\gamma}, \quad |B_z| \gtrsim 10^{-5}, \tag{2.81}$$

---

[8] Which you can do, of course, on the course's web page, and for a few $\mathrm{R_m}$ values, moreover...

2.6 The CP Flow and Fast Dynamo Action

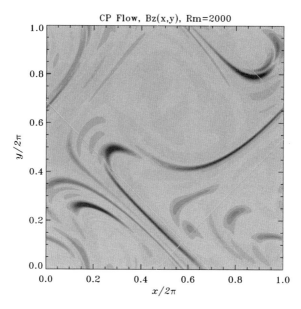

Fig. 2.18 Snapshot of the z-component of the magnetic field in the $[x, y]$ plane, for a CP flow solution with $R_m = 2000$ and $k = 0.57$, in the asymptotic regime $t \gg \tau_c$. The color scale codes the real part of the z-component of $\boldsymbol{b}(x, y, t)$ (*gray-to-blue* is negative, *gray-to-red* positive). The *green straight lines* indicate the separatrix surfaces of the underlying pattern of flow cells, and are no longer fixed in space due to the precession of the flow cells (see Eqs. (2.78)–(2.80)). This is a strongly time-dependent solution, exhibiting overall exponential growth of the magnetic field.

spanning over four orders of magnitude in field strength, and with $\gamma \simeq 0.75$ here. This indicates that strong fields are still far more likely to be detected than if the magnetic field was simply a normally-distributed random variable (for example). The fact that the power law index γ is smaller than unity means that the *largest* local field strength found in the domain will always dominate the computation of the spatially-averaged field strength.

The CP flow dynamo solutions also exhibit *temporal intermittency*; if one sits at one specific point (x, y) in the domain and measures B_z at subsequent time steps, a weak B_z is measured most of the time, and only occasionally are large values detected. Once again the PDF shows a power-law tail with slope flatter than -1 indicating that a temporal average of B_z at one location will always be dominated by the largest B_z measured to date.

Unlike in the Roberts cell, the range of k yielding dynamo action does *not* shift significantly to higher k as R_m is increased, and in the high R_m regime the corresponding maximum growth rate k_{max} does *not* decrease with increasing R_m (as it does in Fig. 2.16). In the CP flow considered here ($A = C = \sqrt{3/2}, \omega = 1, \varepsilon = 1$), $k_{max} \simeq 0.57$, with $s(k_{max}) \simeq 0.3$ for $R_m \gtrsim 10^2$, as shown on Fig. 2.20 (solid line). Figure 2.20 indicates that the CP flow operates as a fast dynamo.

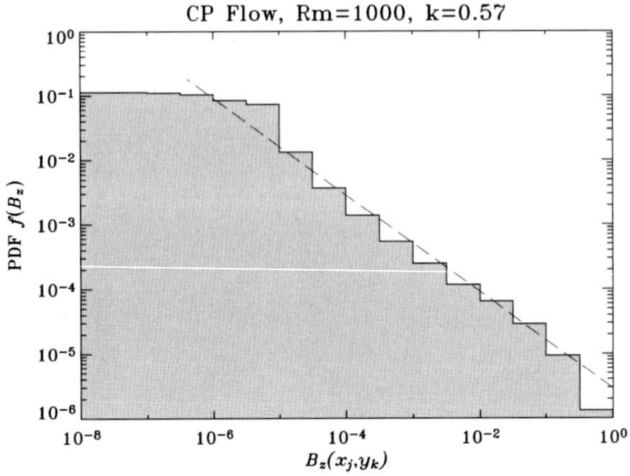

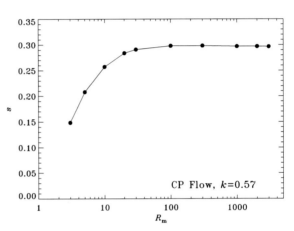

Fig. 2.19 Probability density function for the (unsigned) strength of the z-component of the magnetic field, for a $R_m = 10^3$, $k = 0.57$ CP flow dynamo. The peak field strength has been normalized to a value of unity. Note the power-law tail at large field strength (*straight line* in this log-log plot, with slope ~ -0.75).

Fig. 2.20 Growth rate of $k = 0.57$ CP flow dynamo solutions, plotted as a function of the magnetic Reynolds number (*solid line*). The constancy of the growth rate in the high-R_m regime suggests (but does not strictly prove) that this dynamo is fast.

2.6.2 Fast Dynamo Action and Chaotic Trajectories

Fast dynamo action in the CP flow turns out to be intimately tied to the presence of *chaotic trajectories* in the flow. Their presence (or absence) in a given flow can be quantified in a number of ways, the most straightforward (in principle) being the calculations of the flow's *Lyapunov exponents*. This is another fancy name for a rather simple concept: the rate of exponential divergence of two neighbouring fluid element located at x_1, x_2 at $t = 0$ somewhere in the flow. The Lyapunov exponent λ_L can be (somewhat loosely) defined via

$$\ell(t) = \ell(0) \exp(\lambda_L t), \qquad (2.82)$$

where $\ell \equiv \|x_2 - x_1\|$ is the length of the *tangent vector* between the two fluid elements. Because there are three independent possible directions in 3D space, one can compute *three* distinct Lyapunov exponents at any given point in the flow, and it can be shown that for an incompressible flow their sum is zero. Now, recalling the

2.6 The CP Flow and Fast Dynamo Action

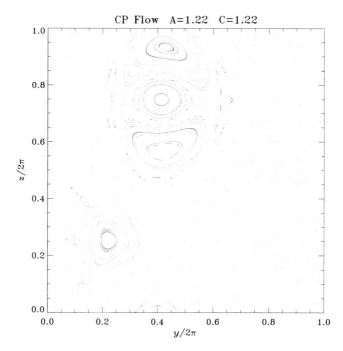

Fig. 2.21 Poincaré section for the CP flow, for $\varepsilon = 1$, $\omega = 1$, and $A = C = \sqrt{3/2}$. The plot is constructed by repeatedly "launching" particles at $z = 0$, $t = 0$, following their trajectories in time, and plotting their (projected) position (modulo 2π) in the $[x, y]$ plane at interval $\Delta t = 2\pi$. The flow is chaotic within the featureless "salt-and-pepper" regions, and integrable in regions threaded by closed curves.

simple flux tube stretching example of Sect. 2.2, exponential divergence of two points located in the same fieldline within the tube clearly implies exponential increase in the tube's length, and therefore, via Eq. (2.19), exponential increase of the magnetic field strength. Theorems have been proven, demonstrating that

1. A smooth flow cannot be a fast dynamo if $\lambda_L = 0$, so that $\lambda_L > 0$, or, equivalently, the existence of chaotic regions in the flow is a necessary (although not sufficient) condition for fast dynamo action;
2. In the limit $R_m \to \infty$, the largest Lyapunov exponent of the flow is an upper bound on the dynamo growth rate.

Proofs of these theorems need not concern us here (if curious see bibliography), but they once again allow us to *rule out* fast dynamo action in many classes of flows.

Calculating a *Poincaré section*, as plotted on Fig. 2.21 for our CP flow, is another very useful way to check for chaotic trajectories in a flow. It is constructed by launching tracer particles at $z = 0$ (and $t = 0$), and following their trajectories as they are carried by the flow. At every 2π time interval, the position of each particle is plotted in the $[x, y]$ plane (modulo 2π in x and y, since most particles leave the original 2π-domain within which they were released as a consequence of cell precession).

Some particles never venture too far away from their starting position in the $[x, y]$ plane. They end up tracing closed curves which, however distorted they may end up looking, identify regions of space where trajectories are integrable. Other particles, on the other hand, never return to their starting position. If one waited long enough, one such particle would eventually come arbitrarily close to all points in the $[x, y]$ plane outside of the integrable regions. The corresponding particle trajectory is said to be *space filling*, and the associated particle motion chaotic. The region of the $[x, y]$ plane defined by the starting positions of all particles with space filling trajectories is called the *chaotic region* of the flow.

2.6.3 Magnetic Flux Versus Magnetic Energy

With the CP flow, we definitely have a pretty good dynamo on our hands. But how are those dynamo solutions to be related to the sun (or other astrophysical bodies)? So far we have concentrated on the magnetic energy as a measure of dynamo action, but in the astrophysical context *magnetic flux* is also important. Consider the following two (related) measures of magnetic flux:

$$\Phi = |\langle \boldsymbol{B} \rangle|, \qquad F = \langle |\boldsymbol{B}| \rangle, \qquad (2.83)$$

where the angular brackets indicate some sort of suitable spatial average over the whole computational domain. The quantity Φ is nothing but the average magnetic flux, while F is the average *unsigned flux*. Under this notation the magnetic energy can be written as $\mathcal{E}_B = \langle \boldsymbol{B}^2 \rangle$.

Consider now the scaling of the two following ratios as a function of the magnetic Reynolds number:

$$\mathcal{R}_1 = \frac{\mathcal{E}_B}{\Phi^2} \propto \mathrm{R}_\mathrm{m}^n, \qquad (2.84)$$

$$\mathcal{R}_2 = \frac{F^2}{\Phi^2} \propto \mathrm{R}_\mathrm{m}^\kappa. \qquad (2.85)$$

A little reflection will reveal that a large value of $\mathcal{R}_1$ indicates that the magnetic field is concentrated in a small total fractional area of the domain, i.e., the *filling factor* is much smaller than unity.[9] The ratio $\mathcal{R}_2$, on the other hand, is indicative of the dynamo's ability to generate a net signed flux. The exponent κ measures the level of *folding* in the solution; large values of κ indicate that while the dynamo may be vigorously producing magnetic flux on small spatial scales, it does so in a manner such that very little *net* flux is being generated on the spatial scale of the computational domain. Figure 2.22 shows the variations with R_m of the two ratios defined above.

---

[9] If you can't figure it out try this: take a magnetic field of strength B_1 crossing a surface area A_1; now consider a more intense magnetic field, of strength $B_2 = 4B_1$, concentrated in one quarter of the area A_1; calculate $\mathcal{E}_B$, Φ, and $\mathcal{R}_1$... get it?

2.6 The CP Flow and Fast Dynamo Action

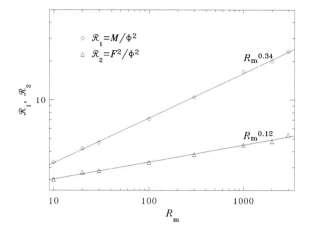

Fig. 2.22 Variations with R_m of the two ratios defined in Eqs. (2.84)–(2.85). Least squares fits (*solid lines*) yield power law exponents $n = 0.34$ and $\kappa = 0.12$.

Least squares fits to the curves yields $n = 0.34$ and $\kappa = 0.12$. Positive values for the exponents κ and n indicate that the CP flow dynamo is relatively inefficient at producing magnetic flux in the high R_m regime, and even less efficient at producing net signed flux. While other flows yielding fast dynamo action lead to different values for these exponents, in general they seem to always turn out positive, with $\kappa < n$, so that the (relative) inability to produce net signed flux seems to be a generic property of fast dynamos in the high-R_m regime.

2.6.4 Fast Dynamo Action in the Nonlinear Regime

We conclude this section by a brief discussion of fast dynamo action in the nonlinear regime. Evidently the exponential growth of the magnetic field will be arrested once the Lorentz force becomes large enough to alter the original CP flow. What might the nature of the backreaction on $\boldsymbol{u}$ look like?

Naively, one might think that the Lorentz force will simply reduce the amplitude of the flow components, leaving the overall geometry of the flow more or less unaffected. That this cannot be the case becomes obvious upon recalling that in the high R_m regimes the eigenfunction is characterized by magnetic structures of typical thickness $\propto R_m^{-1/2}$, while the flow has a typical length scale $\sim 2\pi$ in our dimensionless units. The extreme disparity between these two length scales in the high-R_m regime suggests that the saturation of the dynamo-generated magnetic field will involve alterations of the flow field on small spatial scales, so that a flow very much different from the original CP flow is likely to develop in the nonlinear regime.

That this is indeed what happens was nicely demonstrated some years ago by F. Cattaneo and collaborators (see references in the bibliography), who computed simplified nonlinear solutions of dynamo action in a suitably forced CP flow. They could show that

1. the r.m.s. flow velocity in the nonlinearly saturated regime is comparable to that in the original CP flow;
2. magnetic dissipation actually *decreases* in the nonlinear regime;
3. dynamo action is suppressed by the disappearance of chaotic trajectories in the nonlinearly modified flow.

2.7 Dynamo Action in Turbulent Flows

The Roberts cell and CP flow are arguably more akin to malfunctioning washing machines than any sensible astrophysical object. Nonetheless many things we have learned throughout this chapter do carry over to more realistic circumstances, and in particular to turbulent, thermally-driven convective fluid motions.

As support for this grand sweeping claim, consider Fig. 2.23 herein. It is a snapshot of a numerical simulation of dynamo action in a stratified, thermally-driven turbulent fluid being heated from below and spatially periodic in the horizontal directions. The fluid is contained in a rectangular box of aspect ratio $x : y : z = 10 : 10 : 1$, and we are here looking at the top layer of the simulation box, with the color scale encoding the vertical magnetic field component $B_z(x, y)$. Such thermally-driven turbulent flows in a stratified background have long been known to be characterized by cells of broad upwellings of warm fluid. These cells have a horizontal size set by, among other things, the density scale height within the box. On the other hand, the downwelling of cold fluid needed to satisfy mass conservation end up being concentrated in a network of narrow lanes at the boundaries between adjacent upwelling cells. This asymmetry is due to the vertical pressure and density gradient in the box: rising fluid expands laterally into the lower density layers above, and descending fluid is compressed laterally in the higher density layers below. Near the top of the simulation box, this leads to the concentration of magnetic structures in the downwelling lanes, as they are continuously being swept horizontally away from the centers of upwelling cells, through a form of flux expulsion in fact.

This convectively-driven turbulent flow acts as a vigorous nonlinear fast dynamo, with a ratio of magnetic to kinetic energy of about 20%. This dynamo, much like that arising in the CP flow, produces

1. magnetic fields that are highly intermittent, both spatially and temporally;
2. flux concentrations on scales $\propto R_m^{-1/2}$;
3. little or no mean-field, i.e., signed magnetic flux on a spatial scale comparable to the size of the system.

The fundamental physical link between turbulent convection and the CP flow is the presence of chaotic trajectories in both flows, which leads to the expectation that dynamo action should be possible in convection zones of the sun and stars. Time to move on, then, to the solar magnetic field and its underlying dynamo mechanism(s).

2.7 Dynamo Action in Turbulent Flows

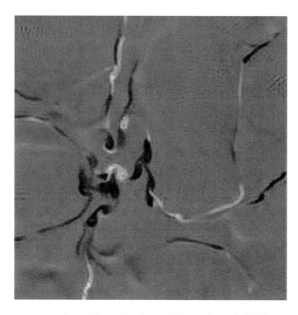

Fig. 2.23 Closeup on a snapshot of the top "horizontal" $[x, y]$ plane of a MHD numerical simulation of thermally-driven stratified turbulent convection in a box of aspect ratio $x : y : z = 10 : 10 : 1$, at a viscous Reynolds number of 245 and $R_m = 1225$. The simulation uses a pseudo-spectral spatial discretization scheme, with 1024 collocation points in the x and y directions, and 97 in z. The color scale encodes the vertical (z) component of the magnetic field (*orange-to-yellow* is positive B_z, *orange-to-blue* negative). Numerical simulation results kindly provided by F. Cattaneo, University of Chicago.

Bibliography

The first detailed discussions of the diffusive decay of large-scale magnetic fields in astrophysical bodies are due to

Cowling, T. G.: 1945, *On the Sun's general magnetic field*, Mon. Not. Roy. Astron. Soc., **105**, 166–174

Wrubel, M. H.: 1952, *On the decay of a primeval stellar magnetic field*, Astrophys. J., **116**, 291–289

On the numerical solution of algebraic eigenvalue problems in general, and on inverse iteration in particular, see

Golub, G. H., & van Loan, C. F.: 1996, *Matrix Computations*, 3rd edn., The John Hopkins University Press, Sects. 7.6.1 and 7.7.8

Press, W. H., Teukolsky, S. A., Vetterling, W. T., & Flannery, B. P.: 1992, *Numerical Recipes. The Art of Scientific Computing*, 2nd edn., Cambridge University Press, Sect. 11.7

Chapter 19 of Press et al. also contains a good introduction to the art of PDE discretization by finite differences, including the simple explicit schemes used to compute the various solutions discussed in Sects. 2.3, 2.5 and 2.6.

Among the numerous books discussing the calculation of macroscopic transport coefficients such as viscosity and electrical conductivity, starting from a microscopic point of view, I still like the following old classic:

Spitzer, Jr., L.: 1962, *Physics of Fully Ionized Gases*, 2nd edn., Wiley Interscience

On the Stretch-Twist-Fold dynamo, see

Vainshtein, S. I., & Zel'dovich, Y. B.: 1972, *Reviews of topical problems: origin of magnetic fields in astrophysics (turbulent "dynamo" mechanisms)*. Soviet Physics Uspekhi, **15**, 159–172

Moffatt, H. K., & Proctor, M. R. E.: 1985, *Topological constraints associated with fast dynamo action*, J. Fluid Mech., **154**, 493–507

Magnetic flux expulsion from regions of closed streamlines is discussed in many textbooks dealing with magnetohydrodynamics. Analytical solutions for some specific cellular flows can be found for example in

Moffatt, H. K.: 1978, *Magnetic Field Generation in Electrically Conducting Fluids*, Cambridge University Press

Parker, E. N.: 1979, *Cosmical Magnetic Fields: Their Origin and their Activity*, Clarendon Press, Chap. 16

but for the first and last word on this topic, you should consult

Weiss, N. O.: 1966, *The expulsion of magnetic flux by eddies*. Proc. Roy. Soc. London Ser., A **293**, 310–328

Rhines, P. B., & Young, W. R.: 1983, *How rapidly is a passive scalar mixed within closed streamlines?*, J. Fluid Mech., **133**, 133–145

The Rhines & Young paper contains clean analytical examples of the two successive dissipative phases with characteristic timescales proportional to $R^{1/3}$ and R_m, as discussed in Sect. 2.3.4. Flux expulsion is of course not restricted to 2D flows; for a nice example in 3D see

Galloway, D. J., & Proctor, M. R. E.: 1983, *The kinematics of hexagonal magnetoconvection*, Geophys. Astrophys. Fluid Dyn., **24**, 109–136

A good recent review of the the current state of planetary magnetic field observations and models can be found in:

Stanley, S., & Glatzmaier, G. A.: 2010, *Dynamo models for planets other than Earth*, Space Sci. Rev., **152**, 617–649

On the possible symmetrization of the Saturnian magnetic field by deep-seated differential rotation, and associated dynamo-based explanations, see

Kirk, R. L., & Stevenson, D. J.: 1987, *Hydromagnetic constraints on deep zonal flow in the giant planets*, Astrophys. J., **316**, 836–846

Christensen, U. R., & Wicht, J.: 2008, *Models of magnetic field generation in partly stable planetary cores: Applications to Mercury and Saturn*, Icarus, **196**, 16–34

Stanley, S.: 2010, *A dynamo model for axisymmetrizing Saturn's magnetic field*, Geophys. Res. Lett., **37**, L05201

An insightful discussion of the symmetrization process in more general terms is that of

Rädler, K.-H.: 1986, *On the effect of differential rotation on axisymmetric and nonaxisymmetric magnetic fields of cosmic bodies*, in *Plasma Astrophysics*, Guyenne, T. D., & Zeleny, L. M., eds., ESA Special Publication, 251, Joint Varenna-Abastumani International School and, Workshop, 569–574

On anti-dynamo theorems, see

Cowling, T. G.: 1933, *The magnetic field of sunspots*, Mon. Not. Roy. Astron. Soc., **94**, 39–48

Bullard, E., & Gellman, H.: 1954, *Homogeneous dynamos and terrestrial magnetism*, Phil. Trans. Roy. Soc. London Ser., A, **247**, 213–278

Zel'dovich, Y. B.: 1956, *The magnetic field in the two-dimensional motion of a conducting turbulent fluid*, Zhurnal Eksperimentalnoi i Teoreticheskoi Fiziki, **31**, 154–156, in Russian

Zel'dovich, Y. B., & Ruzmaikin, A. A.: 1980, *The magnetic field in conducting fluid in two-dimensional motion*, Zhurnal Eksperimentalnoi i Teoreticheskoi Fiziki, **78**, 980–986, in Russian

as well as pages 113-ff and 538-ff, respectively, of the books by Moffatt and Parker listed above. English translations of the last two papers are also reprinted in

Ostriker, J. P., Barenblatt, G. I., & Sunyaev, R. A.: 1992, *Selected Works of Yakov Borisovich Zeldovich. Volume 1: Chemical Physics and Hydrodynamics.*, Princeton University Press

The mathematical aspects of fast dynamo theory are discussed at length in the book of

Childress, S., & Gilbert, A. D.: 1995, *Stretch, Twist, Fold: The Fast Dynamo*, Springer

although the reader preferring a shorter introduction might want to first work through the three following review articles:

Childress, S.: 1992, *Fast dynamo theory*, in *Topological Aspects of the Dynamics of Fluids and Plasmas*, Moffatt, H., Zaslavsky, G. M., Compte, P., & Tabor, M., eds., Kluwer Academic, 111–147

Soward, A. M.: 1994, *Fast dynamos*, in *Lectures on Solar and Planetary Dynamos*, Proctor, M. R. E., & Gilbert, A. D., eds., Cambridge University Press, 181–218

Gilbert, A. D.: 2003, *Dynamo theory*, in *Handbook of Mathematical Fluid Dynamics*, ed. S. Frielander & D. Serre, Vol. 2, Elsevier Science, 355–441

On the Roberts cell dynamo, see Chapter 5 of the Childress & Gilbert book cited above, as well as

Roberts, G. O.: 1972, *Dynamo action of fluid motions with two-dimensional periodicity*, Phil. Trans. Roy. Soc. London Ser., A **271**, 411–454

Soward, A. M.: 1987, *Fast dynamo action in a steady flow*, J. Fluid Mech., **180**, 267–295

The scaling relations given by Eqs. (2.68)–(2.69) are derived in the second of these papers. On fast dynamo action in the CP flow, begin with:

Galloway, D. J., & Proctor, M. R. E.: 1992, *Numerical calculations of fast dynamos in smooth velocity fields with realistic diffusion*, Nature **356**, 691–693

Ponty, Y., Pouquet, A., & Sulem, P. L.: 1995, *Dynamos in weakly chaotic two-dimensional flows*, Geophys. Astrophys. Fluid Dyn., **79**, 239–257

Cattaneo, F., Kim, E.-J., Proctor, M., & Tao, L.: 1995, *Fluctuations in quasi-two-dimensional fast dynamos*, Phys. Rev. Lett., **75**, 1522–1525

The discussion in Sect. 2.6.3 follows closely that in the Cattaneo et al. (1995) paper, and the discussion of nonlinear effects in the CP flow is largely inspired from

Cattaneo, F., Hughes, D. W., & Kim, E.-J.: 1996, *Suppression of chaos in a simplified nonlinear dynamo model*, Phys. Rev. Lett., **76**, 2057–2060

but on this topic see also

Mininni, P. D., Ponty, Y., Montgomery, D. C., Pinton, J.-F., Politano, H., & Pouquet, A.: 2005, *Dynamo regimes with a nonhelical forcing*, Astrophys. J., **626**, 853–863

The mathematically inclined reader wishing to delve deeper into the theorems for fast dynamo action mentioned in Sect. 2.6 will get a solid and character building workout out of

Vishik, M. M.: 1989, *Magnetic field generation by the motion of a highly conducting fluid*, Geophys. Astrophys. Fluid Dyn., **48**, 151–167

Klapper, I., & Young, L. S.: 1995, *Rigorous bounds on the fast dynamo growth rate involving topological entropy*, Comm. Math. Phys., **173**, 623–646

On small-scale dynamo action in three-dimensional thermally-driven convective turbulence, see, e.g.,

Cattaneo, F.: 1999, *On the origin of magnetic fields in the quiet photosphere*, Astrophys. J. Lett., **515**, L39–L42

Cattaneo, F., Emonet, T., & Weiss, N.: 2003, *On the interaction between convection and magnetic fields*, Astrophys. J., **588**, 1183–1198

Stein, R. F., & Nordlund, Å.: 2006, *Solar small-scale magnetoconvection*, Astrophys. J., **642**, 1246–1255

Vögler, A., & Schüssler, M.: 2007, *A solar surface dynamo*, Astron. & Astrophys., **465**, L43–L46

Chapter 3
Dynamo Models of the Solar Cycle

> *Was einmal gedacht wurde,*
> *kann nicht mehr zurückgenommen werden.*
>
> Friedrich Dürrenmatt
> Die Physiker (1962)

The time has now come to put together everything (well... almost) we have learned so far to construct dynamo models for solar and stellar magnetic fields. In this and the following chapter we concentrate on the sun, for which the amount of available observational data constrains dynamo models to a degree much greater than for other stars. Dynamo action in stars other than the sun will be considered in Chap. 5, using solar dynamo models as skyhooks.

We begin (Sect. 3.1) by briefly reviewing the basic properties of the solar magnetic cycle, which are to be (hopefully) reproduced by the (relatively) simple dynamo models to be constructed in the remainder of this chapter. These dynamo models all share the shearing of a poloidal field by differential rotation (Sect. 2.2.4) as a source of toroidal field, and all invoke some sort of enhanced, "turbulent" magnetic diffusivity in the solar convective envelope (more on that very shortly!). They differ primarily in the choice they make regarding the physical mechanism responsible for the regeneration of the poloidal magnetic component.

The first stop in our modelling journey is a statistical-physical theory known as *mean-field electrodynamics*, which will allow us to construct (relatively) simple dynamo models in which the poloidal field is produced through the inductive action of convective turbulence, as described by mean-field electrodynamics (Sect. 3.2). We then look into what currently stands as their main "competitors", namely solar cycle models based on poloidal field regeneration by the surface decay of active regions, more succinctly known as Babcock–Leighton models (Sect. 3.3). We then turn to cycle models relying on various hydrodynamical or MHD instabilities, which can under certain circumstances act as sources of poloidal magnetic fields (Sect. 3.4). We carry on with an overview of the current state of affairs with regards to investigations of the solar dynamo problem through large-scale MHD simulations of turbulent convection in a thick, stratified rotating shell (Sect. 3.5). The chapter closes (Sect. 3.6). with a brief look at some results from local MHD simulations of photospheric and subsurface convection, and what they can teach us regarding the possible multiplicity of dynamo mechanisms in the sun and stars.

3.1 The Solar Magnetic Field

3.1.1 Sunspots and the Photospheric Magnetic Field

The sun is the first astronomical object (Earth excluded) in which a magnetic field was detected, through the epoch-making work of George Ellery Hale (1868–1938) and collaborators. In 1907–1908, by measuring the Zeeman splitting in magnetically sensitive lines in the spectra of *sunspots* and detecting the polarization of the split spectral components (see Fig. 3.1), Hale provided the first unambiguous and quantitative demonstration that sunspots are the seat of strong magnetic fields. Not only was this the first detection of a magnetic field outside the Earth, but the inferred magnetic field strength, 0.3 T, turned out a few thousand times greater than the Earth's own magnetic field. It was subsequently realized that the Lorentz force associated with such strong magnetic fields would also impede convective energy transport from below, and therefore lead naturally to the lower temperatures observed within the sunspots, as compared to the surrounding photosphere.

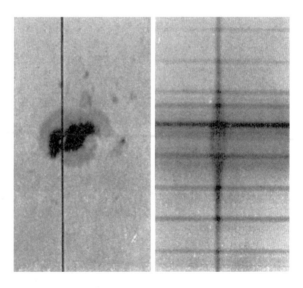

Fig. 3.1 The magnetically-induced Zeeman splitting in the spectrum of a sunspot. The *vertical dark line* on the *left* image is the slit having produced the vertical stack of spectra on the *right* image (with wavelength running horizontally). Reproduced from the 1919 paper by Hale et al. by permission of the AAS.

The solar surface magnetic field outside of sunspots, although of much weaker strength, is accessible to direct observation, usually by measuring Zeeman

3.1 The Solar Magnetic Field

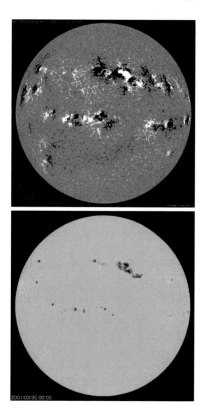

Fig. 3.2 Full disk line-of-sight magnetogram (*top*) and continuum intensity image (*bottom*) of the solar photosphere, both taken on 30 March 2001 by the MDI instrument onboard the SOHO satellite. The sun's rotation axis is vertical on both images. Public domain images downloaded from the SOHO mission website: http://sohowww.nascom.nasa.gov.

broadening of spectral lines, or the degree of linear or circular polarisation of light emitted from the solar photosphere. The first magnetic maps, or *magnetograms*, of the solar disk were obtained in the late 1950s by the father-and-son team of Harold D. Babcock (1882–1968) and Horace W. Babcock (1912–2003), and were little more than photographs of a few dozen horizontally stacked scans of the solar disk displayed on an oscilloscope. Figure 3.2 (top) is a modern equivalent in pixel form, with the gray scale coding the strength of the normal component of the magnetic field (mid-level gray, $|\boldsymbol{B}| \lesssim 1$ mT; going to white for positive normal field, and to black for negative, peaking around 0.4T in both cases). Comparison with a continuum image (bottom) reveals that the stronger magnetic fields coincide with sunspots, but hefty fields of a few 10^{-2} tesla can be found within and around groups of sunspots. Away from these "magnetically active regions", the magnetic field is weaker, concentrated into clumps that collectively make up a spatially fragmented *magnetic network* distributed evenly over the whole surface. Sunspots and active regions, in contrast, are restricted to heliographic latitudes $\lesssim 40°$, and their number waxes and wanes on an 11 yr cycle, about which we'll have a lot more to say in the next chapter.

Now then, to sum up: far from taking the form of a large-scale, smooth diffuse field as on the Earth, the solar photospheric magnetic field is very fragmented and topologically complex, and shows up concentrated in small magnetized regions, separated by field-free plasma. This dichotomy persists down to the smallest spatial scales that can be resolved with current observational techniques. It owes much to the fact that the outer 30% in radius of the sun is a fluid in a strongly turbulent state. Observations at high time-cadence and spatial resolutions of the solar small-scale magnetic field have shown that the associated photospheric magnetic flux is replenished on an hourly timescale, commensurate with the convective turnover time immediately below the photosphere. Such observations have also shown that the magnetic flux of small-scale magnetic structures visible at the solar surface distributed according to a power-law spanning over 5 orders of magnitude in flux, a remarkable instance of scale invariance. These observations offer strong support to a turbulent dynamo-based explanation for the solar small-scale magnetic field, of the type considered in the preceding chapter, away from active regions at least, although other explanations are also possible (more on these in Sect. 3.6 further below).

From here onwards, we focus mostly on the *large-scale* solar magnetic field, by which we mean the part of the sun's magnetic field spatially organized on scales commensurate with the solar radius. While it may not be immediately obvious on Fig. 3.2, sunspots provide one of the better tracers of this large-scale magnetic component.

3.1.2 Hale's Polarity Laws

Hale and his collaborators did much more than just measure magnetic fields in sunspots. Through painstaking observations and analyses spanning nearly two decades, they went on to demonstrate the existence of a number of regularities in the magnetic fields of sunspots, now known as *Hale's polarity laws*. Having noted early on that large sunspots often appear grouped in pairs of opposite magnetic polarities, they could show that:

1. At any given time, the polarity of the leading spots (with respect to the direction of solar rotation) of sunspot pairs is the same in a given solar hemisphere;
2. At any given time, the polarity of the leading spots of sunspot pairs is opposite in the N and S hemispheres;
3. Sunspot polarities reverse in each hemisphere from one 11-yr sunspot cycle to the next.

This polarity ordering is fairly easy to discern on the magnetogram of Fig. 3.2. The most straightforward interpretation of this common opposite polarity grouping is that we are seeing the surface manifestation of a large-scale toroidal field residing somewhere below the photosphere, having risen upwards and pierced the photosphere in the form of a so-called "Ω-loop" (see Fig. 3.3), its intersection with the photosphere producing sunspot pairs of opposite polarities. If this is the case, and if the magnetic flux ropes have not suffered too much twisting about the axis defined by the trajectory of their apex, then the sign of the deep-seated toroidal component B_ϕ is then given by the magnetic polarity of the trailing sunspots. This picture of sunspot pairs, taken

3.1 The Solar Magnetic Field

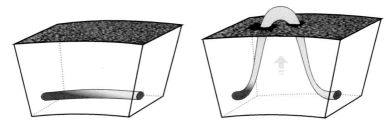

Fig. 3.3 Schematic representation of a sunspot pair as the manifestation of an underlying toroidal flux rope having risen through the photosphere as an "Ω-loop". At *left*, the flux tube lies in the azimuthal direction, before destabilisation and buoyant rise through the photosphere (at *right*). The magnetic fields impedes convective energy transport, so that cooling leads to a collapse of the magnetic field into two sunspots of opposite polarities. Diagram kindly provided by D. Passos.

in conjunction with Hale's polarity laws, therefore indicate that the sun's internal toroidal field is antisymmetric about the equator and reverses polarity from one sunspot cycle to the next.

Another pattern uncovered by Hale and collaborators is that the line segment joining two members of a sunspot pair tends to show a systematic tilt angle (γ) with respect to the East–West direction, the sunspot farther ahead (in the direction of solar rotation) being closer to the equator. Although there exists considerable variations in observed tilt angles, statistically the magnitude of the tilt increases with increasing heliocentric latitude. This is known as *Joy's law*. Least-squares fits to observations yield a parametric representation of the form:

$$\sin \gamma = 0.5 \cos(\theta) , \qquad (3.1)$$

where θ is the usual polar angle. This pattern plays an important role in some of the solar cycle models to be considered later in this chapter. This is because the existence of a finite, systematic tilt implies a net dipole moment, which can contribute to the solar poloidal field.

The hemispheric antisymmetry evidenced by Hale's polarity laws can be readily produced by the shearing of a large-scale poloidal field by a differential rotation symmetric about the equatorial plane, exactly as we modeled already in Sect. 2.2.4. The very existence of Hale's polarity laws thus suggests the presence of a large-scale poloidal component to the solar magnetic field; its detection was beyond the capability of Hale's instruments, but later observations clearly established its existence, and its close connection to the internal magnetic field through the solar magnetic cycle.

3.1.3 The Magnetic Cycle

Figure 3.4 is a synoptic (time-latitude) diagram of the longitudinally-averaged photospheric radial magnetic field component, covering three sunspot cycles. Such a diagram is constructed by averaging magnetograms (like the one on Fig. 3.2) in longitude over each successive solar rotation, and stacking side-by-side the resulting latitudinal distribution of ϕ-averaged magnetic field to form a temporal sequence. The most immediately striking global patterns apparent on Fig. 3.4 are certainly the cyclic variations on a ~ 22 yr period, accompanying polarity reversals, and the (anti)symmetry about the solar equator.

The magnetic signal present within the latitudinal band extending $30°$ or so on either side of the equator is the magnetographic imprint of sunspots. Their strong magnetic fields (~ 0.1 T) almost average out on such synoptic diagram, because, as already noted, they tend to appear in close pairs of opposite magnetic polarities with comparable (unsigned) magnetic flux. At the beginning of a sunspot cycle (e.g., 1976, 1986, 1996 on Fig. 3.4), sunspots are observed at relatively high ($\sim 40°$) heliocentric latitudes, but emerge at lower and lower latitudes as the cycle proceeds, until at the end of the cycle they are seen mostly near the equator, at which time spots announcing the onset of the next cycle begin to appear again at $\sim 40°$ latitude. This results in the so-called "butterfly diagram" of sunspot distribution, about which we'll have more to say in the following chapter. Cycle maximum (as measured by sunspot number) occurs about midway along each butterfly, when sunspot coverage is maximal at about $15°$ latitude, here 1980, 1991 and 2002.

At high heliocentric latitude ($\gtrsim 50°$) the synoptic magnetograms are dominated by a well-defined dipole component, with strength $\sim 10^{-3}$ T, showing a clean pattern of polarity changes occurring at or near sunspot maximum. For example, during

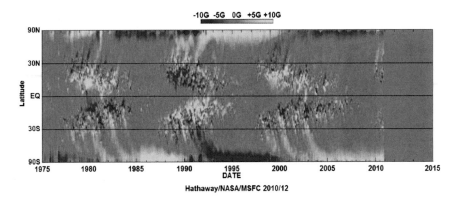

Fig. 3.4 A synoptic magnetogram covering the last three sunspot cycles. The radial component of the sun's magnetic field is azimuthally averaged over a solar rotation, and the resulting latitudinal strips stacked one against the other in the form of a time-latitude diagram. Recall that $1\text{T} \equiv 10^4$ gauss. Data and graphics courtesy of David Hathaway, NASA/MSFC, http://solarscience.msfc.nasa.gov/images/magbfly.jpg.

the 1976–1986 cycle the toroidal field was negative in the N-hemisphere, and the Northern polar field reversed from positive to negative magnetic polarity; taken at face value, Fig. 3.4 then indicates that the high latitude poloidal field lags the toroidal field by a phase interval $\Delta\varphi \simeq \pi/2$.

At mid-latitudes the most prominent feature is a fairly regular poleward drift of magnetic fields originating in sunspot latitudes, presumably released there by the decay of sunspot and active regions. It is quite possible that this poleward transport of magnetic flux from active region belts contributes to the polarity reversal of the polar fields.

A $\sim 10^{-3}$ T polar field pervading a polar cap of $\sim 30°$ angular width, as on Fig. 3.4, adds up to a poloidal magnetic flux of $\sim 10^{14}$ Wb. The total unsigned flux emerging in active regions, taken to be representative of the solar internal toroidal magnetic component, adds up to a few 10^{17} Wb over a full sunspot cycle. This is usually taken to indicate that the solar internal magnetic field is dominated by its toroidal magnetic component.

We will return to sunspots and their spatiotemporal variations in numbers in Sect. 4.1, when we consider the origin of fluctuations in the solar cycle. For the time being we will just concentrate on what they tell us about the strength of the sun's *internal* magnetic field.

3.1.4 Sunspots as Tracers of the Sun's Internal Magnetic Field

In translating the cartoon of Fig. 3.3 into a quantitative physical model, we have a number of issues that need to be clarified. The first is to identify the region(s) of the solar interior from which the flux ropes originate. The magnetic pressure ($\propto \boldsymbol{B}^2$) within a strongly magnetized flux tube leads to a density deficit in order to reach pressure equilibrium with the surrounding plasma. The resulting upward buoyancy force can bring the tube to the surface, which is good and needed, but it turns out that for tubes located within the bulk of the convection zone the rise time is far too short to allow field amplification to a level commensurate with observed sunspot field strengths. This has led to the conclusion that the solar magnetic field is stored— maybe even produced—not in the convective envelope proper, but rather immediately below it, within the tachocline.

Considerable efforts have gone into making models of the storage, destabilization and buoyant rise of thin magnetic flux tubes through the solar convective envelope (see bibliography at the end of this chapter). In most cases, flux tubes are treated as structureless, flux-carrying material lines—the so-called thin flux tube approximation—and so these kinds of calculations cannot properly take into account the interaction of the tube with the surrounding turbulent fluid motions. With this caveat in mind, thin flux tube modelling has produced the following two important results:

1. The flux ropes rise essentially radially if they have a field strength in excess of $B \gtrsim 6\text{--}10\,\text{T}$; otherwise the Coriolis force deflects the rising flux tubes to high latitudes;
2. The flux ropes emerge without any significant tilt for $B \gtrsim 20\,\text{T}$, and with tilts compatible with Joy's law for fields strengths in the range 6–16 T.

The basic physical mechanism underlying these two remarkable results is the same: if the rise time of the flux ropes is of the order of the solar rotation period, the Coriolis force has an important influence. It is the Coriolis force that, upon acting on the internal flow developing along the length of the flux rope during its rise, gives rise to the twist that, at emergence, manifests itself as Joy's law. If the field is strong enough for the rise time to be much shorter than the rotation period, then the rising flux rope does not "feel" the rotation, rises radially, and emerges without a tilt. If on the other hand the magnetic field is too weak, the Coriolis force deflects the rising flux rope on a trajectory running parallel to the rotation axis, resulting in emergence at high heliographic latitudes.

Now, this is great stuff: the observed emergence of sunspots at low heliocentric latitudes puts a *lower* limit on the strength of the participating flux ropes; Joy's law, on the other hand, translate into an *upper* limit on the field strength. One concludes that the sunspot-forming toroidal flux ropes must have magnetic field strengths in the rather narrow range

$$6 \lesssim B \lesssim 16\,\text{T}. \tag{3.2}$$

While some level of field amplification is likely during the (ill-understood) process of flux tube formation from the spatially diffuse large-scale magnetic field produced by the dynamo, these modelling results are usually taken to indicate that the large-scale toroidal magnetic field at or below the base of the convective envelope, where stability analyses indicate sunspots-forming toroidal flux rope are formed and stored, must have a strength in the range of a few tenths to a few tesla. By most estimates, the associated magnetic energy density is at least comparable, and perhaps quite a bit larger than the kinetic energy density of the turbulent fluid motions driving dynamo action.

3.1.5 A Solar Dynamo Shopping List

To close this brief overview, let's now collect a short list of fundamental observational features that a physical model of the solar large-scale magnetic field should reproduce (anything related to amplitude fluctuation being deferred to Chap. 4):

1. A large-scale magnetic field, axisymmetric to a good approximation and antisymmetric about the solar equatorial plane;
2. A cyclic variation of this large-scale magnetic field, characterized by polarity reversals with a ~ 20 yr oscillation period;

3. An internal toroidal field of strength ~ 0.1–1 T, concentrated at low solar latitudes ($\lesssim 45°$, say), and migrating equatorward in the course of the cycle with minimal spatiotemporal overlap between successive cycles;
4. A large-scale surface poloidal field of a few 10^{-3} T, migrating poleward in the course of the cycle, and reversing polarity at sunspot maximum.

These properties do not square well with fast dynamo action in turbulent flow; in particular, the sun's large-scale magnetic field component is characterized by a substantial signed (hemispheric) magnetic flux, for which something else than fast dynamo action is needed. It turns out that the turbulent nature of the flows in the solar convective envelope can still do the trick, but to examine this we will need to adopt a statistical approach to turbulence and to the associated flow-field interactions.

3.2 Mean-Field Dynamo Models

The "toy" dynamo flows considered in Sects. 2.5 and 2.6 exemplified the fact that high-R_m turbulent flows can be quite effective at producing a lot of small-scale magnetic fields, where "small-scales" is roughly $R_m^{-1/2}$ times the length scale of the flow. At the solar surface, the latter is around $\sim 10^6$ m and $R_m \sim 10^8$ (for granulation), which yields very small scales indeed, ~ 100 m! So, at some level, the small-scale magnetic fields on the sun and stars are already taken care of. It turns out that under certain conditions, solar/stellar convective turbulence can also produce magnetic fields with a mean component building up on large spatial scales. These *mean-field dynamo models* remain arguably the most "popular" descriptive models for dynamo action in the sun and stars, but also in planetary metallic cores, stellar accretion disks, and even galactic disks. Accordingly, we will look into the formulation of these models at some depth.[1]

3.2.1 Mean-Field Electrodynamics

The fundamental idea on which mean field theory rests is the *two scales approach*, which consists of a decomposition of the field variables into mean and fluctuating parts. This process naturally implies that an averaging procedure can meaningfully be defined. The derivation of mean field theory can proceed equally from the choice of space averages, time averages or ensemble averages. In the context of axisymmetric dynamo models, longitudinal averages impose themselves rather naturally. For the time being let's just define our averaging operator as:

---

[1] Sections 3.2.1 through 3.2.6 are to a large extent adapted from class notes written by Thomas J. Bogdan for the graduate class APAS7500 we co-taught in 1997 at the University of Colorado at Boulder.

$$\langle A \rangle = \frac{1}{\lambda^3} \int_V A \, d\mathbf{x} \, . \tag{3.3}$$

We also assume that the velocity and magnetic field can be decomposed into a mean and fluctuating part so that

$$\mathbf{u} = \langle \mathbf{u} \rangle + \mathbf{u}' \quad \text{and} \quad \mathbf{B} = \langle \mathbf{B} \rangle + \mathbf{B}' \, . \tag{3.4}$$

The decomposition (3.4) makes sense provided $\langle \mathbf{u}' \rangle = \langle \mathbf{B}' \rangle = 0$. This is *not* a linearization, in that it involves no assumption regarding the relative magnitudes of the mean and fluctuating parts. The physical interpretation of (3.4) is as follows. The velocity and magnetic fields are characterized by a slowly varying component, $\langle \mathbf{u} \rangle$ and $\langle \mathbf{B} \rangle$, which vary on the characteristic large scale L, plus rapidly fluctuating parts, $\mathbf{u}'$ and $\mathbf{B}'$, which vary on the much smaller scale ℓ. The volume averages are computed over some intermediate scale λ such that

$$\ell \ll \lambda \ll L \, . \tag{3.5}$$

Whenever (3.5) is satisfied we say that we have a "good" scale separation.

The objective of mean field theory is to produce a closed set of equations for the mean quantities. Substituting (3.4) into the induction equation (1.59), and averaging, we the obtain the equation for the mean magnetic field

$$\boxed{\frac{\partial \langle \mathbf{B} \rangle}{\partial t} = \nabla \times \left(\langle \mathbf{u} \rangle \times \langle \mathbf{B} \rangle + \boldsymbol{\mathcal{E}} - \eta \nabla \times \langle \mathbf{B} \rangle \right)} \, . \tag{3.6}$$

Subtracting this expression from the full MHD induction equation, obtained by substitution of (3.4) into (1.59) *without* applying the averaging operator, yields the following evolutionary equation for the fluctuating part of the magnetic field:

$$\frac{\partial \mathbf{B}'}{\partial t} = \nabla \times \left(\langle \mathbf{u} \rangle \times \mathbf{B}' + \mathbf{u}' \times \langle \mathbf{B} \rangle + \mathbf{G} - \eta \nabla \times \mathbf{B}' \right), \tag{3.7}$$

where

$$\boldsymbol{\mathcal{E}} = \langle \mathbf{u}' \times \mathbf{B}' \rangle, \quad \text{and} \quad \mathbf{G} = \mathbf{u}' \times \mathbf{B}' - \langle \mathbf{u}' \times \mathbf{B}' \rangle \, . \tag{3.8}$$

The important thing is that (3.6) now contains a source term, $\boldsymbol{\mathcal{E}}$, associated with the average of products of fluctuations, which in general does *not* vanish upon averaging even though $\mathbf{u}'$ and $\mathbf{B}'$ individually do. The term $\boldsymbol{\mathcal{E}}$, which is called the *mean electromotive force*, or emf for short, plays a central role in this theory.

Now, the whole point of the mean-field procedure is to avoid having to deal explicitly with the small scales, so we do not want to be integrating Eq. (3.7) explicitly. But then we have a closure problem: Eq. (3.6) is a 3-component vector equation, for the six components of $\langle \mathbf{B} \rangle$ and $\mathbf{B}'$ (leaving the flow out of the picture for the moment).

3.2 Mean-Field Dynamo Models

Therefore it is clear that to solve (3.6), $\mathcal{E}$ must be expressed as some function of $\langle u \rangle$ and $\langle B \rangle$.

In order to obtain the desired expression, we note that (3.7) is a *linear* equation for B' with the term $\nabla \times (u' \times \langle B \rangle)$ acting as a source. There must therefore exist a *linear* relationship between B and B', and hence, one between B and $\langle u' \times B' \rangle$. The latter relationship can be expressed formally by the following series

$$\mathcal{E}_i = \alpha_{ij} \langle B \rangle_j + \beta_{ijk} \partial_k \langle B \rangle_j + \gamma_{ijkl} \partial_j \partial_k \langle B \rangle_l + \cdots, \tag{3.9}$$

where the tensorial coefficients, α, β, γ, and so forth must depend on $\langle u \rangle$, on what we might loosely term the *statistics* of the turbulent velocity fluctuations, u', and perhaps on the diffusivity η—but *not* on $\langle B \rangle$. In this sense, Eqs. (3.6) and (3.9) constitute a closed set of equations for the evolution of $\langle B \rangle$. The convergence of the series representation provided by Eq. (3.9) can be anticipated in those cases where the good separation of scales applies. For in these cases each successive derivative in Eq. (3.9) is smaller than the previous one by approximately a factor of $\ell/L \ll 1$. With any luck, we may expect Eq. (3.9) to be dominated by the first few terms.

3.2.2 The α-Effect

We have already remarked that $\mathcal{E}$ in (3.6) acts as a source term for the mean field. It is instructive to examine the contributions to $\mathcal{E}$ deriving from the individual terms in the expansion (3.9). The first contribution is associated with the second-rank tensor, α_{ij}, thus

$$\mathcal{E}_i^{(1)} = \alpha_{ij} \langle B \rangle_j . \tag{3.10}$$

The first thing to note is that α_{ij} must be a pseudo–tensor since it establishes a linear relationship between a polar vector—the mean emf, and an axial vector—the mean magnetic field. We can divide α_{ij} into its symmetric and antisymmetric parts, thus[2]

$$\alpha_{ij} = \alpha_{ij}^s - \varepsilon_{ijk} a_k , \tag{3.11}$$

where $2a_k = -\varepsilon_{ijk}\alpha_{ij}$. From (3.10) we have

$$\mathcal{E}_i^{(1)} = \alpha_{ij}^s \langle B \rangle_j + (a \times \langle B \rangle)_i . \tag{3.12}$$

---

[2] Here, ε_{ijk} is the Levi–Civita tensor density, also known as the unit alternating tensor, and has the values $\varepsilon_{ijk} = 0$ when i, j, k are not all different, $\varepsilon_{ijk} = +1$ or -1 when i, j, k are all different and in cyclic, or acyclic order, respectively. A particularly useful formula is (Einstein summation over repeated indices in force):

$$\varepsilon_{ijk}\varepsilon_{klm} = \delta_{il}\delta_{jm} - \delta_{im}\delta_{jl} ,$$

where δ_{ij} is the Kronecker–delta, and has the value $\delta_{ij} = 0$ if i, j are different, and $\delta_{ij}=1$ when $i=j$.

The effect of the antisymmetric part is to provide an additional advective velocity (not in general solenoidal), so that the effective mean velocity becomes $\langle \boldsymbol{u} \rangle + \boldsymbol{a}$. It results in *turbulent pumping* of the large-scale magnetic component. The nature of the symmetric part is most easily illustrated in the case when $\boldsymbol{u}'$ is an *isotropic* random field.[3] Then $\boldsymbol{a}$ is zero, α_{ij} must be an isotropic tensor of the form $\alpha_{ij} = \alpha \delta_{ij}$, and (3.12) reduces to

$$\mathcal{E}^{(1)} = \alpha \langle \boldsymbol{B} \rangle. \tag{3.13}$$

Using Ohm's law, this component of the emf is found to generate a contribution to the mean current of the form

$$\boldsymbol{j}^{(1)} = \alpha \, \sigma_e \langle \boldsymbol{B} \rangle, \tag{3.14}$$

where σ_e is the electrical conductivity. For nonzero α, Eq. (3.14) implies the appearance of a mean current everywhere *parallel* to the mean magnetic field—the so-called α–effect. This is in sharp contrast to the more conventional case where the induced current $\sigma_e (\boldsymbol{u} \times \boldsymbol{B})$ is *perpendicular* to the magnetic field. We are used to thinking of electrical currents being the source of magnetic fields (think of the Biot–Savart law, or the pre-Maxwellian form of Ampère's law); but a mechanically forced magnetic field can become a source of electrical current. That's really what induction is all about.

In the context of axisymmetric large-scale astrophysical magnetic fields, the importance of the α-effect is immediately apparent. We recall from our deliberations in Sect. 2.2.4 that a toroidal field could be generated from a poloidal one by differential rotation (velocity shear). The α-effect makes it possible to drive a mean toroidal current parallel to the mean toroidal field, which, in turn will regenerate a poloidal field thereby closing the dynamo loop.

To appreciate the physical nature of the α-effect we pause to examine the original 1955 physical picture put forth by E.N. Parker. We define a cyclonic event to be the rising of a fluid element associated with a definite twist, say anticlockwise when seen from below (see Fig. 3.5). In spherical geometry, we then consider the effect of many such events, distributed randomly in longitude and time, on an initially purely toroidal fieldline. Each cyclonic event creates an elemental loop of field with an associated current distribution that will have a component parallel to the initial field if the angle of rotation is less than π and antiparallel if it is greater. By assuming that the individual events are short lived we can rule out rotations of more than 2π. It is clear that the combined effect of many such events is to give rise to a net current with a component along $\langle \boldsymbol{B} \rangle$.

---

[3] Throughout the rest of this chapter, we will have cause to repeatedly refer to the statistical properties of the turbulent velocity field. In order to avoid confusion we state the following definitions: a (random) field is *stationary* if its probability density function (pdf) is time independent, it is *homogeneous* if its pdf is independent of position, it is *isotropic* if its pdf is independent of orientation (or equivalently, invariant under rotations), and it is *reflectionally symmetric* if its pdf is invariant under parity reversal. We should note that isotropy and reflectional symmetry are taken here to be distinct properties, although this protocol is not universally accepted.

3.2 Mean-Field Dynamo Models

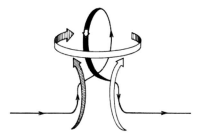

Fig. 3.5 A sketch of a magnetic line of force entrained by a cyclonic, rising fluid element in the frozen-in limit. Note that the resulting cyclonic loop can be viewed as resulting from an element of electric current flowing parallel to the original, uniform magnetic field. From Parker (1970), Fig. 1, by permission of the AAS.

An important property of α is its pseudoscalar nature, i.e., α changes sign under parity transformations. This implies that α can be nonzero only if the statistics of $\boldsymbol{u}'$ lacks reflectional symmetry. In other words, the velocity field must have a definite chirality. For example, on Fig. 3.5 there is a definite relationship between vertical displacements and sense of twist, set by the Coriolis force. This is similar to the Roberts cell flow, where the z-component changed sign in step with the sense of rotation in contiguous flow cells. In general the lack of reflectional symmetry of the fluid velocity manifests itself through a nonzero value of the mean fluid helicity $\langle \boldsymbol{u}' \cdot (\nabla \times \boldsymbol{u}') \rangle$, itself a pseudo scalar. The Roberts Cell and CP flows introduced in the preceding chapter are two examples of flows lacking reflectional symmetry. As we shall presently see, there is an important relation between fluid helicity and the α-effect.

It is important to establish those cases in which α can rigorously be computed from knowledge of $\boldsymbol{u}'$. Not counting methods based on the direct numerical solutions of the induction equation, there are two distinct ways to proceed. In both cases the success of the approach depends on some simplification of Eq. (3.7). In one case the term $\boldsymbol{G}$ is neglected leading to the so-called first order smoothing approximation (FOSA). In the other, the term $\eta \nabla^2 \boldsymbol{B}'$ is neglected, leading to the Lagrangian approximation. The two approaches are complementary in the sense that the former is applicable (for most physically relevant circumstances) when the diffusivity is large and the latter when it is small. Even these two most severe simplifying assumptions do not exactly lead to simple mathematics, and to add insult to injury, the parameter regimes for which they are expected to hold do not square well with what we think we know about solar interior conditions. The closest we can get to the sun and stars, in a tractable manner, is the so-called Second-order correlation approximation (SOCA), which neglects cross-correlations between the different velocity components but retains the possibility that the intensity of turbulence itself can vary with position. Under this assumption of near-isotropy, we then have

$$\langle u'_j u'_k \rangle = \frac{1}{3} \langle (\boldsymbol{u}')^2 \rangle \delta_{jk} \,. \tag{3.15}$$

This leads to a simple diagonal form for the α tensor:

$$\boxed{\alpha = -\frac{1}{3}\tau_c \langle \mathbf{u}' \cdot (\nabla \times \mathbf{u}') \rangle \quad [\text{m s}^{-1}]}, \tag{3.16}$$

where τ_c is the correlation time for the turbulent flow. Equation (3.16) tells us that the α-effect is a direct function of the helicity of the turbulent component of the flow; think back of Parker's picture of twisted magnetic fieldlines (Fig. 3.5) and convince yourself that this is indeed how it should be for the "cartoon" to work.

If one assumes that the (mild) inhomogeneity arises from the stratification, the (mild) break of reflectional symmetry from the Coriolis force, and the lifetime of turbulent eddies is commensurate with their turnover time, then Eq. (3.16) can be brought to the form:

$$\alpha = -\frac{1}{3}\tau_c^2 (u')^2 \mathbf{\Omega} \cdot \nabla \ln(\varrho u'), \tag{3.17}$$

where $u' = \sqrt{\langle \mathbf{u}'^2 \rangle}$ is the local r.m.s. turbulent velocity, and $\mathbf{\Omega}$ is the angular velocity vector. With the turbulent velocity increasing outwards through the convective envelope faster than the density decreases, Eq. (3.17) would "predict" an α-effect varying as $\cos\theta$ and positive (negative) in the solar Northern (Southern) hemisphere. Such expression can be validated through MHD numerical simulations of turbulent flows including an externally-imposed weak magnetic field, and from the simulation statistics one can then compute the α-tensor components by appropriate averaging.[4] There has been many such simulations, which, almost surprisingly, have corroborated the expressions obtained from SOCA.

The key parameter is the so-called Coriolis number, defined as the ratio of rotation period to convective turnover time:

$$\text{Co} = 2\Omega \tau_c, \tag{3.18}$$

equivalent to the inverse of the Rossby number of common usage in atmospheric sciences. The Coriolis number is a dimensionless measure of the ability of the Coriolis force to break the mirror symmetry of convective turbulence. Estimates for this quantity in the sun, with τ_c taken from mixing length theory, yield $\text{Co} \ll 1$ in the outer convection envelope, up to $\text{Co} \sim 1\text{--}10$ at its base. For $\text{Co} \sim 1$, the $\alpha_{\phi\phi}$ component of the α-tensor, which is the term responsible for poloidal field regeneration in axisymmetric mean-field models, does turn out positive in the bulk of the convection zone, with a $\sim \cos\theta$ latitudinal dependency. At larger rotation rate, the peak in $\alpha_{\phi\phi}$ is displaced from the pole to lower latitudes, reaching $\sim 30°$ at $\text{Co} \sim 10$. These simulations also produce a sign change in all components of the α-tensor at the very base of the convective envelope, with the region of negative $\alpha_{\phi\phi}$ growing in size as Co increases from 1 to 10.

---

[4] See the papers by Ossendrijver et al. (2001) and Käpylä et al. (2006) cited in the bibliography, on which the foregoing discussion is based.

3.2 Mean-Field Dynamo Models

The above expressions for the α-coefficients are predicated on the small-scale field $\boldsymbol{B}'$ being much weaker than the mean-field $\langle\boldsymbol{B}\rangle$, a situation expected to hold only in the $R_m \ll 1$ regime, or if the coherence time of the turbulent flow is much smaller than its turnover time. The first condition is the regime entirely opposite to that expected in solar/stellar interiors, while the second is at best marginally satisfied. High-R_m MHD turbulence simulations indicate that in this regime one has in fact $\boldsymbol{B}' \gtrsim \langle\boldsymbol{B}\rangle$, and that Eq. (3.16) should be replaced by:

$$\alpha = -\frac{1}{3}\tau_c\left(\langle\boldsymbol{u}'\cdot(\nabla\times\boldsymbol{u}')\rangle - \frac{1}{\varrho}\langle\boldsymbol{J}'\cdot\boldsymbol{B}'\rangle\right), \quad (3.19)$$

with $\boldsymbol{J}' = (\nabla\times\boldsymbol{B}')/\mu_0$. Notice that the second term on the RHS, corresponding to the current helicity associated with the small-scale magnetic field, has a sign opposite to that of the kinetic helicity. This says once again, in essence, that the Lorentz force opposes the twisting of the large-scale magnetic field by the turbulent flow. This impact of current helicity on the α-effect represents a potentially powerful quenching mechanism for the α-effect, a topic we shall revisit further below.

3.2.3 Turbulent Pumping

The non-diagonal part of the α tensor provides a contribution to the turbulent emf taking the form of a non-solenoidal advective velocity (second term on RHS of Eq. (3.12)). This is emphatically *not* a real flow, in the sense that it acts only on the large-scale magnetic component, and originates with the turbulent emf. Turbulent pumping can also be measured in numerical simulations, which indicate that the predominant effect is a downward pumping driven by the stratification, with magnetic fields being expelled from the high-diffusivity regions to the low diffusivity regions. In the presence of rotation, turbulent pumping also takes place in the latitudinal direction, with a velocity reaching values of the order of a few meters per second at high rotation rates (Co = 10).

Although turbulent pumping is seldom explicitly included in the simple mean-field dynamo models to be discussed presently, its impact on dynamo action in the sun and solar-type stars is likely important; this is because it can offset flux loss through magnetic buoyancy, and favors accumulation of magnetic fields in the tachocline, where the large shear and low magnetic diffusivity are conducive to the production of strong toroidal flux rope-like structures, believed to give rise to sunspots following their destabilization, buoyant rise through the convection zone and surface emergence.

3.2.4 The Turbulent Diffusivity

We now turn to the next term in the expansion (3.9), namely

$$\mathcal{E}_i^{(2)} = \beta_{ijk}\partial_k \langle B \rangle_j. \tag{3.20}$$

The physical interpretation of the third-rank pseudotensor, β_{ijk}, is again most easily gained when $\boldsymbol{u}'$ is isotropic, in which case $\beta_{ijk} = \beta\varepsilon_{ijk}$, where β is a scalar, and so we have

$$\nabla \times \boldsymbol{\mathcal{E}}^{(2)} = \nabla \times \left(-\beta \nabla \times \langle \boldsymbol{B} \rangle \right). \tag{3.21}$$

We recognize the scalar β as an additional contribution to the effective diffusivity of $\langle \boldsymbol{B} \rangle$, which thus becomes $\eta_e \equiv \eta + \beta$. In cases where $\beta \gg \eta$ one refers to $\eta_e \approx \beta$ as the *turbulent diffusivity*. For homogeneous and isotropic turbulence, it can be formally related to the intensity of turbulence as

$$\boxed{\beta = \frac{1}{3}\tau_c \langle (\boldsymbol{u}')^2 \rangle \ [\mathrm{m^2 s^{-1}}]}, \tag{3.22}$$

where τ_c is once again the correlation time of the turbulent flow.[5] Equation (3.22) states that the turbulent diffusivity is more efficient when the turbulence is more vigorous, which makes intuitive sense since, as shown in Sect. 2.3, the rate at which magnetic fieldlines are folded and expelled from (turbulent) flow cells increases with the velocity of the flow.

Simple mixing length models of solar convection suggest $u' \sim 10\,\mathrm{m\,s^{-1}}$ and $\tau_c \sim 1$ month at the base of the convection zone ($r/R \sim 0.7$), which then leads to $\beta \sim 10^8\,\mathrm{m^2\,s^{-1}}$. This is very much larger than the ordinary magnetic diffusivity $\eta \sim 1\,\mathrm{m^2 s^{-1}}$, so that we indeed expect $\beta \gg \eta$. This is why, back in the previous chapter (Sects. 2.1.4 and 2.2.4), whenever trying to model the "real" sun we made use of a magnetic diffusivity profile characterized by a sharp increase when going from the radiative core to the overlying convective envelope (viz. Eq. (2.16) and dash-dotted line on Fig. 2.2). Note also that the magnetic diffusion time (1.63) obtained using the above numerical estimate β for the solar convection zone (with $\ell \sim 0.3R$) is now $\sim 10\,\mathrm{yr}$, which is commensurate to the solar cycle period, and suggests that (turbulent) dissipation can be expected to play an important role in solar cycle models.

---

[5] This expression still holds under SOCA, in which case β becomes a function of position in the flow.

3.2.5 The Mean-Field Dynamo Equations

In summary, our heuristic treatment of mean-field electrodynamics has led us to an evolution equation for the large-scale magnetic field, $\langle \boldsymbol{B} \rangle$, which takes account of coherences between fluctuation-fluctuation interactions of the small-scale turbulent magnetic and velocity fields. For homogeneous, stationary, and isotropic velocity turbulence, this equation assume the particularly elegant and physically intuitive form

$$\frac{\partial \langle \boldsymbol{B} \rangle}{\partial t} = \nabla \times \left(\langle \boldsymbol{u} \rangle \times \langle \boldsymbol{B} \rangle + \alpha \langle \boldsymbol{B} \rangle - \beta \nabla \times \langle \boldsymbol{B} \rangle \right), \tag{3.23}$$

which, according to SOCA, should remain valid in the case of mildly-inhomogeneous, mildly anisotropic turbulence as well, with α and β then given by Eqs. (3.16) and (3.22). The fluctuation-fluctuation interactions enter this equation through the electromotive force described by the α-effect, incarnating what we earlier called constructive folding, and the turbulent diffusion of the mean magnetic field accounted for by β, tantamount to destructive folding. In principle, these coefficients can be calculated from the lowest-order statistics of the turbulent flow, namely the spatial distribution of turbulent intensity, as measured by $\langle (\boldsymbol{u}')^2 \rangle$.

The fact remains that more often than not, and certainly in all mean-field dynamo models to be considered in what follows, the mean-field coefficients α and β will be chosen a priori, although we will take care to embody in these choices what we have learned from our brief excursion into mean-field theory. Consequently, the resulting dynamo models will have a descriptive, rather than predictive value. We will be picking numerically "reasonable" turbulent dynamo coefficient that "do the right thing" for the sun, and see how the resulting models behave as we change other aspects of the model, or, later on, apply them to stars other than the sun. Yet, as the following example will show, we can still learn a lot from mean-field electrodynamics, even though we have foregone strict physical determinism.

3.2.6 Dynamo Waves

As discussed already in Sect. 3.1 the shape of the sunspot butterfly diagram suggests that the sunspot-forming deep-seated toroidal magnetic flux system migrates equatorward in the course of the cycle. It turns out that this remarkable pattern can arise naturally in the context of cycle models based on mean-field electrodynamics.

Consider a local cartesian coordinate system oriented so that the direction y corresponds to an ignorable coordinate ($\partial/\partial y = 0$), which we associate with the azimuthal direction in an axisymmetric spherical system, and with x and z mapping onto the latitudinal and radial directions, respectively. Consider now the action of a spatially constant α-effect acting in conjunction with a vertically-sheared flow:

$$\langle \boldsymbol{u} \rangle = \Omega z \, \hat{\boldsymbol{e}}_y, \tag{3.24}$$

where Ω is a constant [units: s^{-1}]. We shall further assume that the mean-field coefficients α [units: m s^{-1}] and $\eta_e = \beta + \eta$ [units: m^2 s^{-1}] are constant. The cartesian equivalent of Eq. (2.2) is now

$$\langle \boldsymbol{B}\rangle(x, z, t) = \nabla \times (A(x, z, t)\hat{\boldsymbol{e}}_y) + B(x, z, t)\hat{\boldsymbol{e}}_y \,. \tag{3.25}$$

Substitution of Eqs. (3.24) and (3.25) into our mean-field induction equation (3.23) leads to

$$\frac{\partial A}{\partial t} - \eta_e \left(\frac{\partial^2 A}{\partial x^2} + \frac{\partial^2 A}{\partial z^2}\right) = \alpha B \,, \tag{3.26}$$

$$\frac{\partial B}{\partial t} - \eta_e \left(\frac{\partial^2 B}{\partial x^2} + \frac{\partial^2 B}{\partial z^2}\right) = \Omega \frac{\partial A}{\partial x} - \alpha \left(\frac{\partial^2 A}{\partial x^2} + \frac{\partial^2 A}{\partial z^2}\right) \,. \tag{3.27}$$

The two terms on the RHS of this equation parameterize the α-effect and the Ω-effect. Recall that the Ω-effect describes the generation of a toroidal magnetic field by the shearing of a poloidal field (as in Sect. 2.2.4). The (mean-field) α-effect accounts for the regeneration of *both* poloidal and toroidal magnetic fields due to the chirality, or handedness, of the turbulent flow field. These two terms offer the possibility of dynamo action overcoming the magnetic diffusion term which resides on the LHS of this equation.

Equations (3.26) and (3.27) are again PDEs with constant coefficients. We can therefore seek elementary plane-wave solutions of the form

$$\begin{bmatrix} A(x, z, t) \\ B(x, z, t) \end{bmatrix} = \begin{bmatrix} a \\ b \end{bmatrix} \exp[\lambda t + \mathrm{i}k(z \cos \vartheta + x \sin \vartheta)] \,. \tag{3.28}$$

We may assume that $k \geq 0$ and $0 \leq \vartheta \leq 2\pi$ are prescribed (real) parameters, where the latter sets the orientation of the wave vector in the $[x, z]$ plane. If Eq. (3.28) is substituted into Eqs. (3.26)–(3.27), the requirement that there be nontrivial eigenvectors leads to the dispersion relation:

$$\left(\lambda + \eta_e k^2\right)^2 = \alpha k \left(\alpha k + \mathrm{i}\Omega \sin \vartheta\right) \,. \tag{3.29}$$

This is a quadratic (complex) polynomial in λ, with the two solutions:

$$\lambda_\pm = -\eta_e k^2 \pm \sqrt{\frac{|\alpha|k}{2}} \left\{\left(\sqrt{\Omega^2 \sin^2 \vartheta + \alpha^2 k^2} + |\alpha|k\right)^{\frac{1}{2}}\right.$$
$$\left. + \mathrm{i}\,\mathrm{sign}(\Omega\alpha \sin \vartheta) \left(\sqrt{\Omega^2 \sin^2 \vartheta + \alpha^2 k^2} - |\alpha|k\right)^{\frac{1}{2}}\right\}, \tag{3.30}$$

with proper care exerted in extracting the square root of complex number using the standard algebraic formulae. The λ_- solution can only produce a disturbance which decays with the passage of time, so our hope rests on the λ_+ root, with dynamo

3.2 Mean-Field Dynamo Models

action occurring when $\mathrm{Re}(\lambda_+) > 0$. Examination of Eq. (3.30) indicates that an exponentially growing *dynamo wave* is obtained when $0 < k < k_\star$, where the critical wavenumber $k_\star$ is one of the (six) roots of the equation,

$$k_\star^6 - \frac{\alpha^2}{\eta_e^2} k_\star^4 - \frac{\alpha^2 \Omega^2}{4\eta_e^4} \sin^2 \vartheta = 0. \tag{3.31}$$

If $k_\star \to 0$ then the "window" for dynamo action disappears. This occurs when $\alpha \to 0$, in agreement with Cowling's theorem. From a physical perspective it makes a good deal of sense that the dynamo window inhabits the small-wavenumber, large-wavelength, end of the range of possible parameters. Clearly dynamo waves with rapid spatial fluctuations are susceptible to severe damping due to the enhanced diffusivity $\eta_e \approx \beta$. On the other hand, if the spatial variations of $\langle A \rangle$ are too large, then there is very little $\langle B \rangle$ for the α-effect to work on, and so the dynamo process again stalls as $k \to 0$.

Equation (3.31) can be solved exactly as a cubic equation for $\zeta \equiv k_\star^2$, but for our purposes it is sufficient to simply estimate $k_\star$ by inspection of Eq. (3.31) in the limiting cases of "strong" shear, usually most relevant to dynamo action in the sun and stars:

$$k_\star \approx \left[\frac{|\alpha \Omega \sin \vartheta|}{2\eta_e^2}\right]^{\frac{1}{3}}, \quad |\alpha| \ll \sqrt{\eta_e |\Omega \sin \vartheta|}. \tag{3.32}$$

We use the word "wave" to describe these exponentially growing solutions of the mean field equations, because it is clear from Eq. (3.30) that $\mathrm{Im}(\lambda_+) \neq 0$. Note also that the direction of propagation clearly depends upon the sign of the product of α and Ω, and that the largest growth rate will occur for $\vartheta = \pi/2$, i.e., wave propagating in the "latitudinal" x-direction, which is a most excellent first step towards reproducing the sunspot butterfly diagram!

3.2.7 The Axisymmetric Mean-Field Dynamo Equations

We now proceed to reformulate the mean-field induction equation (3.23) into a form suitable for axisymmetric large-scale magnetic fields pervading a sphere of electrically conducting fluid. We proceed as we did way back in Sect. 2.4, which is to express the poloidal field as the curl of a toroidal vector potential, and restrict the large-scale flow to the axisymmetric forms given by Eq. (2.59), with the magnetic diffusivity restricted to vary at most only with r. It will also prove convenient to express the resulting equations in nondimensional form.

Toward this end we opt to scale all lengths in terms of R, and time in terms of the diffusion time $\tau = R^2/\eta_e$ based on the (turbulent) diffusivity in the convective envelope, which we assume to be provided by the (scalar) β-term of mean-field electrodynamics. Henceforth dropping the averaging brackets for notational simplicity, the poloidal/toroidal separation procedure applied to the mean-field dynamo equation (3.23) now leads to

$$\frac{\partial A}{\partial t} = \eta \left(\nabla^2 - \frac{1}{\varpi^2} \right) A - \frac{R_m}{\varpi} \boldsymbol{u}_p \cdot \nabla(\varpi A) + C_\alpha \alpha B, \qquad (3.33)$$

$$\frac{\partial B}{\partial t} = \eta \left(\nabla^2 - \frac{1}{\varpi^2} \right) B + \frac{1}{\varpi} \left(\frac{d\eta}{dr} \right) \frac{\partial(\varpi B)}{\partial r} - R_m \varpi \nabla \cdot \left(\frac{B}{\varpi} \boldsymbol{u}_p \right)$$
$$+ C_\Omega \varpi (\nabla \times A\hat{\boldsymbol{e}}_\phi) \cdot (\nabla \Omega) + C_\alpha \hat{\boldsymbol{e}}_\phi \cdot \nabla \times (\alpha \nabla \times (A\hat{\boldsymbol{e}}_\phi)), \quad (3.34)$$

where the following three nondimensional numbers have materialized:

$$C_\alpha = \frac{\alpha_e R}{\eta_e}, \qquad (3.35)$$

$$C_\Omega = \frac{\Omega_e R^2}{\eta_e}, \qquad (3.36)$$

$$R_m = \frac{u_e R}{\eta_e}, \qquad (3.37)$$

with α_e (dimension m s^{-1}), u_e (dimension m s^{-1}) and Ω_e (dimension s^{-1}) as reference values for the α-effect, meridional flow and differential rotation, respectively.

Remember that the functionals α, η, $\boldsymbol{u}_p$ and Ω are hereafter dimensionless. The quantities C_α and C_Ω are *dynamo numbers*, measuring the importance of inductive versus diffusive effects on the RHS of Eqs. (3.33) and (3.34). The third dimensionless number, R_m, is a magnetic Reynolds number, which here measures the relative importance of advection (by meridional circulation) versus diffusion in the transport of A and B in meridional planes. For simplicity of notation, we continue to use η for the total magnetic diffusivity, retaining the possibility of variation with depth and with the understanding that within the convective envelope this now includes the (dominant) contribution from the β-term of mean-field theory.

Equations (3.33) and (3.34) will hereafter be referred to as the *dynamo equations* (rather than the technically preferable but cumbersome "axisymmetric mean-field dynamo equations"). Structurally, they only differ from Eqs. (2.61) to (2.62) by the presence of not one but two new source terms on the RHS, both associated with the α-effect. The appearance of this term in Eq. (3.33) is crucial, since this is what allows us to evade Cowling's theorem. Acting in conjunction with the new α-effect term in Eq. (3.34), it makes dynamo action possible in the absence of a large-scale shear, i.e., with $\nabla \Omega = 0$ in Eq. (3.34). Such dynamos are known as α^2 dynamos, and regenerate both the poloidal and toroidal magnetic fields entirely via the inductive action of small-scale turbulence. Traditionally, dynamo action in planetary cores has been assumed to belong to this variety (at least from the point of view of mean-field theory).

3.2 Mean-Field Dynamo Models

Another possibility is that the shearing terms entirely dominates over the α-effect term, in which case the latter is altogether dropped out of Eq. (3.34). This leads to the $\alpha\Omega$ dynamo model, which is believed to be most appropriate to the sun and solar-type stars. Finally, retaining both source terms in Eq. (3.34) defines, you guessed it I hope, the $\alpha^2\Omega$ dynamo model. This has received comparatively little attention in the context of solar/stellar dynamos, since (simple) a priori estimates of the dynamo numbers C_α and C_Ω usually yield $C_\alpha/C_\Omega \ll 1$; caution is however warranted if dynamo action takes place in a thin shell, in which case the α-term can still dominate toroidal field production.

In general, solutions are sought in a meridional plane of a sphere of radius R, and as with the diffusive problem of Sect. 2.1, are matched to a potential field in the exterior ($r/R > 1$), and regularity requires that $A(r,0) = A(r,\pi) = 0$ and $B(r,0) = B(r,\pi) = 0$ be imposed on the symmetry axis. In practice it is often useful to solve explicitly for modes having odd and even symmetry with respect to the equatorial plane. To do so, one simply solves the dynamo equations in a meridional *quadrant*, and imposes the following boundary conditions along the equatorial plane. For a dipole-like antisymmetric mode,

$$\frac{\partial A(r,\pi/2)}{\partial \theta} = 0, \qquad B(r,\pi/2) = 0, \qquad \text{[Antisymmetric]}, \qquad (3.38)$$

while for symmetric (quadrupole-like) modes one sets instead

$$A(r,\pi/2) = 0, \qquad \frac{\partial B(r,\pi/2)}{\partial \theta} = 0, \qquad \text{[Symmetric]}. \qquad (3.39)$$

We are now ready, if not to rock, at least to roll...

3.2.8 Linear $\alpha\Omega$ Dynamo Solutions

In constructing mean-field dynamos for the sun, it has been a common procedure to neglect meridional circulation, on the grounds that it is a very weak flow (but more on this further below), and to adopt the $\alpha\Omega$ model formulation, on the grounds that with $R \simeq 7 \times 10^8$ m, $\Omega_e \sim 10^{-6}$ rad s^{-1}, and $\alpha_e \sim 1$ m s^{-1}, one finds $C_\alpha/C_\Omega \sim 10^{-3}$, independently of the assumed (and poorly constrained) value for the turbulent diffusivity. We also restrict the models to the kinematic regime, i.e., all flow fields posed a priori and deemed steady ($\partial/\partial t = 0$). Equations (3.33) and (3.34) then reduce to the so-called $\alpha\Omega$ dynamo equations:

$$\frac{\partial A}{\partial t} = \eta \left(\nabla^2 - \frac{1}{\varpi^2} \right) A + C_\alpha \alpha B, \qquad (3.40)$$

$$\frac{\partial B}{\partial t} = \eta \left(\nabla^2 - \frac{1}{\varpi^2} \right) B + C_\Omega \varpi (\nabla \times A\hat{e}_\phi) \cdot (\nabla \Omega) + \frac{1}{\varpi} \frac{d\eta}{dr} \frac{\partial(\varpi B)}{\partial r}, \qquad (3.41)$$

where α, Ω and η are now all dimensionless functions of spatial coordinates, remember. In the spirit of producing a model that is solar-like we use a fixed value $C_\Omega = 2.5 \times 10^4$, obtained assuming $\Omega_e \equiv \Omega_{Eq} \sim 10^{-6}$ rad s^{-1} and $\eta_e = 5 \times 10^7$ m^{2}s^{-1}, which leads to a diffusion time $\tau = R^2/\eta_e \simeq 300$ yr.

In the parameter regime characterizing the strongly turbulent solar convection zone, the strength and spatial variation of the α-effect cannot be computed in any reliable manner from first principles, so this will remain the major unknown of the model. In accordance with the $\alpha\Omega$ approximation of the dynamo equations, we restrict ourselves to cases where $|C_\alpha| \ll C_\Omega$. For the dimensionless functional $\alpha(r, \theta)$ we use an expression of the form

$$\alpha(r, \theta) = f(r)g(\theta), \qquad (3.42)$$

where

$$f(r) = \frac{1}{4}\left[1 + \mathrm{erf}\left(\frac{r - r_c}{w}\right)\right]\left[1 - \mathrm{erf}\left(\frac{r - 0.8}{w}\right)\right]. \qquad (3.43)$$

This combination of error functions concentrates the α-effect in the bottom half of the envelope, and lets it vanish smoothly below, just as the net magnetic diffusivity does (i.e., we again set $r_c/R = 0.7$ and $w/R = 0.05$). Various lines of argument point to an α-effect peaking at the bottom of the convective envelope, since there the convective turnover time is commensurate with the solar rotation period, a most favorable setup for the type of toroidal field twisting at the root of the α-effect. Likewise, the hemispheric dependence of the Coriolis force suggests that the α-effect should be positive in the Northern hemisphere, and change sign across the equator ($\theta = \pi/2$). The "minimal" latitudinal dependency is thus

$$g(\theta) = \cos\theta. \qquad (3.44)$$

The C_α dimensionless number, measuring the strength of the α-effect, is treated as a free parameter of the model. You may be shocked by the fact that we are, in a very cavalier manner, effectively treating the α-effect as a (almost) free-function; this sorry situation is unfortunately the rule rather than the exception in mean-field dynamo modelling.[6]

With α, β and the large-scale flow given, the $\alpha\Omega$ dynamo equations (3.40) and (3.41) become linear in the mean-field $\boldsymbol{B}$. With none of the PDE coefficients depending explicitly on time, one can seek eigensolutions of the form

$$\begin{bmatrix} A(r, \theta, t) \\ B(r, \theta, t) \end{bmatrix} = \begin{bmatrix} a(r, \theta) \\ b(r, \theta) \end{bmatrix} e^{\lambda t}, \qquad (3.45)$$

---

[6] References to some of the more noteworthy exceptions are provided in the bibliography at the end of this chapter.

3.2 Mean-Field Dynamo Models

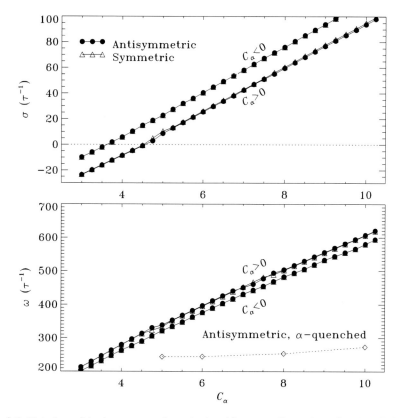

Fig. 3.6 Variations of the dynamo growth rate (*top*) and frequency (*bottom*) as a function of $|C_\alpha|$ in the minimal $\alpha\Omega$ model with solar-like internal differential rotation. Sequences are shown for either positive or negative dynamo number (as labeled), and symmetric (*triangles*) or antisymmetric (*dots*) parity. Modes having $\sigma < 0$ are decaying, and modes with $\sigma > 0$ are exponentially growing. Here modes with A or S parity have very nearly identical eigenvalues. In this model the first mode to reach criticality is the negative C_α mode, for which $D_{\rm crit} = -0.9 \times 10^5$. The positive C_α mode reaches criticality at $D_{\rm crit} = 1.1 \times 10^5$. The *diamonds* on panel **b** correspond to the dynamo frequency measured in a nonlinear version of the same minimal $\alpha\Omega$ model, including algebraic α-quenching, to be discussed in Sect. 3.2.10.

where the amplitudes a and b are in general complex quantities. Substituting Eq. (3.45) into the $\alpha\Omega$ dynamo equations yields a classical linear eigenvalue problem. It will prove convenient to write the eigenvalue explicitly as

$$\lambda = \sigma + i\omega, \qquad (3.46)$$

so that σ is the growth rate and ω the cyclic frequency, both expressed in terms of the inverse diffusion time $\tau^{-1} = \eta_e/R^2$. In a model for the (oscillatory) solar dynamo, we are looking for solutions where $\sigma > 0$ and $\omega \neq 0$.

Armed (and dangerous) with the above model, we plow ahead and solve numerically the $\alpha\Omega$ dynamo equations as a 2D eigenvalue problem. We first produce a sequence of solutions for increasing values of $|C_\alpha|$, holding C_Ω fixed at a its "solar" value 2.5×10^4, Fig. 3.6 shows the variation of the growth rate σ and frequency ω as a function of C_α. Four sequences are shown, corresponding to modes that are either antisymmetric or symmetric with respect to the equatorial plane ("A" and "S" respectively), computed with either positive or negative C_α. For $|C_\alpha|$ smaller than some threshold value, the induction terms make too small a contribution to the RHS of Eq. (3.40), leaving the dissipation terms dominant, so that solutions all have $\sigma < 0$, as per Cowling's theorem. As $|C_\alpha|$ increases, the growth rate eventually reaches $\sigma = 0$. At this point we also have $\omega \neq 0$, so that the corresponding solution oscillates with neither growth of decay of its amplitude. Further increases of $|C_\alpha|$ lead to $\sigma > 0$. We are now finally in the dynamo regime, where a weak initial field is amplified exponentially in time.

Computing similar sequences for the same model but different values of C_Ω soon reveals that the onset of dynamo activity ($\sigma > 0$) is controlled by the *product* of C_α and C_Ω:

$$D \equiv C_\alpha \times C_\Omega = \frac{\alpha_e \Omega_e R^3}{\eta_e^2}. \tag{3.47}$$

The value of D for which $\sigma = 0$ is called the *critical dynamo number* (denoted D_{crit}). This, at least, is similar to what we found for the analytical solution of Sect. 3.2.6. Modes having $\sigma < 0$ are called *subcritical*, and those having $\sigma > 0$ *supercritical*. Note on Fig. 3.6 how little the growth rate and dynamo frequency depend on the assumed solution parity.

Here the first mode to become supercritical is the negative C_α mode, for which $D_{\text{crit}} = -0.9 \times 10^5$, followed shortly by the positive C_α mode ($D_{\text{crit}} = -1.1 \times 10^5$). The dynamo frequency for these critical modes is $\omega \simeq 300$, which corresponds to a full cycle period of ~ 6 yr. This is within a factor of four of the observed full solar cycle period. Once again we should not be too impressed by this, since we have quite a bit of margin of manoeuvre in specifying numerical values for η_e and C_α, and there is no reason to believe that the sun should be exactly at the critical threshold for dynamo action.

Figure 3.7 shows half a cycle of the dynamo solution, in the form of snapshots of the toroidal (color scale) and poloidal (fieldlines) eigenfunctions in a meridional plane, with the rotation/symmetry axis oriented vertically.[7] The four frames are separated by a phase interval $\varphi = \pi/3$, so that panel (d) is identical to (a) except for reversed magnetic polarities in both magnetic components. Such linear eigensolutions leave the *absolute* magnitude of the magnetic field undetermined, but the relative magnitude of the poloidal to toroidal components is found to scale as $\sim |C_\alpha/C_\Omega|$.

The toroidal field peaks in the vicinity of the core–envelope interface, which is not surprising since, in view of Eqs. (2.27) and (2.28), the radial shear is maximal

---

[7] Animation of this solution, and another for its cousin with negative C_α, can be viewed on the course webpage http://obswww.unige.ch/SSAA/sf39/dynamos.

3.2 Mean-Field Dynamo Models

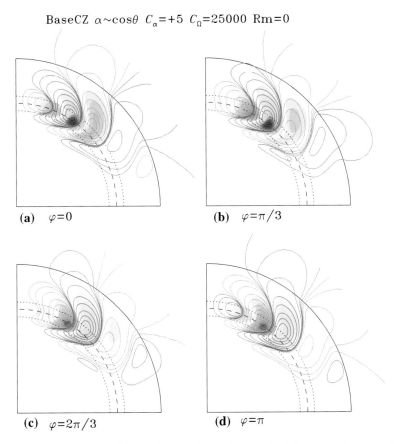

Fig. 3.7 Four snapshots (**a**–**d**) in meridional planes of our minimal linear $\alpha\Omega$ dynamo solution with defining parameters $C_\Omega = 25000$, $\eta_e/\eta_c = 10$, $\eta_e = 5 \times 10^7$ m^2 s^{-1}. With $C_\alpha = +5$, this is a mildly supercritical solution (cf. Fig. 3.6). The toroidal field is plotted as filled contours (*green* to *blue* for negative B, *yellow* to *red* for positive B, normalized to the peak strength and with increments $\Delta B = 0.2$), on which poloidal fieldlines are superimposed (*blue* for clockwise-oriented fieldlines, *orange* for counter-clockwise orientation). The *dashed line* is the core–envelope interface at $r_c/R = 0.7$. The four snapshots shown here cover half a magnetic cycle, i.e., panel **d** is identical to **a** except for reversed magnetic polarities.

there and the magnetic diffusivity and α-effect are undergoing their fastest variation with depth. But why is the amplitude of the dynamo mode vanishing so rapidly below the core–envelope interface? After all, the poloidal and toroidal diffusive eigenmodes investigated in Sect. 2.1 were truly global, and the adopted contrast in magnetic diffusivity between core and envelope should favor stronger fields in the lower diffusivity core. The crucial difference lies with the oscillatory nature of the solution: because the magnetic field produced in the vicinity of the core–envelope interface is oscillating with alternating polarities, its penetration depth in the core is limited by the electromagnetic skin depth $\ell = \sqrt{2\eta_c/\omega}$ (Sect. 2.3), with η_c the core diffusivity. Having assumed $\eta_e = 5 \times 10^7 \, \text{m}^2\text{s}^{-1}$, we have $\eta_c = \eta_e \Delta\eta = 5 \times 10^6 \, \text{m}^2\text{s}^{-1}$. A dimensionless dynamo frequency $\omega \simeq 300$ corresponds to $3 \times 10^{-8} \, \text{s}^{-1}$, so that $\ell/R \simeq 0.026$, quite small indeed.

Careful examination of Fig. 3.7a–d also reveals that the toroidal/poloidal flux systems present in the shear layer first show up at high-latitudes, and then *migrate equatorward* to finally disappear at mid-latitudes in the course of the half-cycle. If you haven't already guessed it: what we are seeing on Fig. 3.7 is the spherical equivalent of the dynamo waves investigated in Sect. 3.2.6 for the cartesian case with uniform α-effect and shear. In more general terms, the dynamo waves travel in a direction s given by

$$s = \alpha \nabla \Omega \times \hat{e}_\phi, \qquad (3.48)$$

i.e., along isocontours of angular velocity. This result is known as the *Parker–Yoshimura sign rule*. Here with a negative $\partial\Omega/\partial r$ in the high-latitude region of the tachocline, a positive α-effect results in an equatorward propagation of the dynamo wave.

3.2.9 Nonlinearities and α-Quenching

Obviously, the exponential growth characterizing supercritical ($\sigma > 0$) linear solutions must stop once the Lorentz force associated with the growing magnetic field becomes dynamically significant for the inductive flow. This magnetic backreaction can show up here in two distinct ways:

1. Reduction of the differential rotation;
2. Reduction of turbulent velocities, and therefore of the α-effect (and perhaps also of the turbulent magnetic diffusivity).

Because the solar surface and internal differential rotation shows very little dependence on the phase of the solar cycle, it has been customary to assume that magnetic backreaction occurs at the level of the α-effect. In the mean-field spirit of *not* solving dynamical equations for the small-scales, it is still a common practice to simply *assume* a dependence of α on B that "does the right thing", namely reducing the α-effect once the magnetic field becomes "strong enough", the latter usually taken to mean when the growing dynamo-generated mean magnetic field reaches a magnitude such that its energy per unit volume is comparable to the kinetic energy of the underlying turbulent fluid motions:

$$\frac{B_{\text{eq}}^2}{2\mu_0} = \frac{\varrho u_t^2}{2} \quad \rightarrow \quad B_{\text{eq}} = u_t \sqrt{\mu_0 \varrho}. \qquad (3.49)$$

This expression defines the *equipartition field strength*, denoted B_{eq}, which varies from ~ 1 T at the base of the solar convective envelope, to ~ 0.1 T in the surface layers. It has become common practice to introduce an ad hoc algebraic nonlinear

3.2 Mean-Field Dynamo Models

quenching of α (and sometimes η_e as well) directly on the mean-toroidal field B by writing:

$$\alpha \to \alpha(B) = \frac{\alpha_0}{1 + (B/B_{\text{eq}})^2}. \tag{3.50}$$

Needless to say, this remains an *extreme* oversimplification of the complex interaction between flow and field that is known to characterize MHD turbulence, but its wide usage in solar dynamo modelling makes it a good choice for the illustrative purpose of this section.

3.2.10 Kinematic $\alpha\Omega$ Models with α-Quenching

With α-quenching included in the poloidal source term, the mean-field $\alpha\Omega$ equations are now nonlinear, and are best solved as an initial-boundary-value problem. The initial condition is an arbitrary seed field of very low amplitude, in the sense that $B \ll B_{\text{eq}}$ everywhere in the domain. Boundary conditions remain the same as for the linear analysis of the preceding section.

Consider again the minimal $\alpha\Omega$ model of Sect. 3.2.8, where the α-effect assumes its simplest possible latitudinal dependency, $\propto \cos\theta$. We use again $C_\Omega = 2.5 \times 10^4$ and positive $C_\alpha \geq 5$, so that the corresponding linear solutions are in the supercritical regime (see Fig. 3.6). With a very weak $\boldsymbol{B}$ as initial condition, early on the model is essentially linear and exponential growth is expected. This is indeed what is observed, as can be seen on Fig. 3.8, showing time series of the total magnetic energy in the simulation domain for increasing values of C_α, all above criticality. Eventually however, B starts to become comparable to B_{eq} in the region where the α-effect operates, leading to a break in exponential growth, and eventual saturation at some constant value of magnetic energy. Evidently, α-quenching is doing what it was designed to do! Note how the saturation energy level increases with increasing C_α, an intuitively satisfying behavior since solutions with larger C_α have a more powerful poloidal source term. The cycle frequency for these solutions is plotted as diamonds on Fig. 3.6b and, unlike in the linear solutions, now shows very little increase with increasing C_α. Moreover, the dynamo frequency of these α-quenched solutions are found to be slightly *smaller* than the frequency of the linear critical mode (here by some 10–15%), a behavior that is typical of these models. Yet the overall form of the dynamo solutions closely resembles that of the linear eigenfunctions plotted on Fig. 3.7. Indeed, the full cycle period is here $P/\tau \simeq 0.027$, which translates into 9 yr for our adopted $\eta_e = 5 \times 10^7 \, \text{m}^2 \, \text{s}^{-1}$, i.e., a little over a factor of two shorter than the real thing. Not bad!

As a solar cycle model, these dynamo solutions do suffer from one obvious problem: magnetic activity is concentrated at too high latitudes (see Fig. 3.7). This is a direct consequence of the assumed $\cos\theta$ dependency for the α-effect. One obvious way to push the dynamo mode towards the equator is to concentrate the α-effect

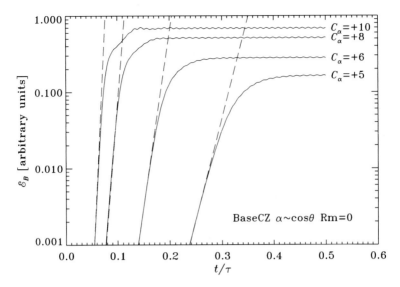

Fig. 3.8 Time series of magnetic energy for a set of $\alpha\Omega$ dynamo solutions using our minimal $\alpha\Omega$ model including algebraic α-quenching, and different values for C_α, as labeled. Magnetic energy is expressed in arbitrary units. The *dashed line* indicates the exponential growth phase characterizing the linear regime.

at low latitude. This is not as ad hoc as one may think, given that the numerical simulation results discussed in Sect. 3.2.2 do indicate that in the high rotation regime (Co $\gtrsim$ 4), the peak in the α-effect is indeed displaced to low latitudes. We therefore proceed using now a latitudinal dependency in $\propto \sin^2\theta \cos\theta$ for the α-effect.

Figure 3.9 shows a selection of three $\alpha\Omega$ dynamo solutions, in the form of time-latitude diagrams of the toroidal field extracted at the core–envelope interface, here $r_c/R = 0.7$. If sunspot-producing toroidal flux ropes form in regions of peak toroidal field strength, and if those ropes rise radially to the surface, then such diagrams are directly comparable to the sunspot butterfly diagram. These three models all have $C_\Omega = 25000$, $|C_\alpha| = 10$, $\Delta\eta = 0.1$, and $\eta_e = 5 \times 10^7$ m^2 s^{-1}. To facilitate comparison between solutions, antisymmetric parity is *imposed* via the boundary condition at the equator. On such diagrams, the latitudinal propagation of dynamo waves shows up as a "tilt" of the flux contours away from the vertical direction.

The first solution, on Fig. 3.9a, is once again our basic solution of Fig. 3.7, with an α-effect varying in $\cos\theta$. The other two use an α-effect varying in $\sin^2\theta \cos\theta$, and so manage to produce dynamo action that materializes in two more or less distinct branches, one associated with the negative radial shear in the high latitude part of the tachocline, the other with the positive shear in the low-latitude tachocline. These two branches propagate in opposite directions, in agreement with the Parker–Yoshimura sign rule, since the α-effect here does not change sign within an hemisphere, but the radial gradient of Ω does.

3.2 Mean-Field Dynamo Models

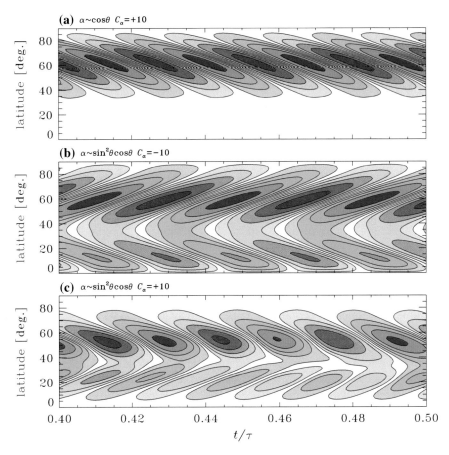

Fig. 3.9 Northern hemisphere time-latitude ("butterfly") diagrams for a selection (**a–c**) of nonlinear $\alpha\Omega$ dynamo solutions including α-quenching, constructed at the depth $r/R = 0.7$ corresponding to the core–envelope interface. Isocontours of the toroidal field are normalized to their peak amplitudes, and plotted for increments $\Delta B/\max(B) = 0.2$, with *yellow*-to-*red* (*green*-to-*blue*) contours corresponding to $B > 0 \, (< 0)$. The assumed latitudinal dependence of the α-effect is given on each panel (**a–c**). Other model ingredients are the same as on Fig. 3.7. Note the co-existence of two distinct cycles in the solution shown on panel **c**, with periods differing by about 25%. Adapted from Living Review of Charbonneau (2010).

It is noteworthy that co-existing dynamo branches, as on Fig. 3.9b, c, can have distinct dynamo periods, which in nonlinearly saturated solutions leads to long-term amplitude modulation. Such modulations are typically not expected in dynamo models where the only nonlinearity present is a simple algebraic quenching formula such as Eq. (3.50). Note that this does not occur for the $C_\alpha < 0$ solution, where both branches propagate away from each other, but share a common latitude of origin and so are phased-locked at the onset (cf. Fig. 3.9b). We are seeing here a first example of potentially distinct dynamo modes interfering with one another, a direct consequence of the complex profile of solar internal differential rotation.

The solution of Fig. 3.9b is characterized by a low-latitude equatorially propagating branch, and a full cycle period of 16 yr, which is getting pretty close to the "target" 22 yr. But again the strong high-latitude, poleward-propagating branch has no counterpart in the sunspot butterfly diagram. This is often summarily dealt with by flatly zeroing out the α-effect at latitudes higher than $\sim 40°$, but this is clearly not a very satisfying approach. Let's try something else instead.

3.2.11 Enters Meridional Circulation: Flux Transport Dynamos

Meridional circulation is unavoidable in turbulent, stratified rotating convection. It basically results from an imbalance between Reynolds stresses and buoyancy forces. The ~ 15 m s^{-1} poleward flow observed at the surface has been detected helioseismically, down to $r/R \simeq 0.85$ without significant departure from the poleward direction, except locally and very close to the surface, in the vicinity of the active region belts. Mass conservation evidently requires an equatorward flow deeper down.

Meridional circulation can bodily transport the dynamo-generated magnetic field (terms $\propto \boldsymbol{u}_p \cdot \nabla$ in Eqs. (2.61) and (2.62)), and therefore, for a (presumably) solar-like equatorward return flow that is vigorous enough, can overpower the Parker–Yoshimura rule and produce equatorward propagation no matter what the sign of the α-effect is. At low circulation speeds, the primary effect is a Doppler shift of the dynamo wave, leading to a small change in the cycle period. The behavioral turnover from dynamo wave-like solutions to circulation-dominated magnetic field transport sets in when the circulation speed in the dynamo region becomes comparable to the propagation speed of the dynamo wave. In the circulation-dominated regime, the cycle period loses sensitivity to the assumed turbulent diffusivity value, and becomes determined primarily by the circulation's turnover time. Solar cycle models achieving equatorward propagation of the deep-seated toroidal field in this manner are often called *flux transport dynamos*.

These properties of dynamo solutions with meridional flows can be cleanly demonstrated in simple $\alpha\Omega$ models using a purely radial shear at the core–envelope interface (see references in bibliography), but with a solar-like differential rotation profile, the situation turns out to be far more complex. Consider for example the three $\alpha\Omega$ dynamo solutions of Fig. 3.9, now recomputed including a meridional flow taking the form of a single cell per meridional quadrant, directed poleward in the outer convective envelope and with the equatorward return flow closing at the core–envelope interface, as illustrated on Fig. 3.10a.[8] As R_m is increased, for the solution of Fig. 3.9a, the dynamo is decaying in $10^2 \lesssim R_m \lesssim 600$, and then kicks in again at $R_m \simeq 800$ with a double-branched structure in its butterfly diagram. The negative-C_α solution (Fig. 3.9b), on the other hand transits to a steady mode

---

[8] An animation of this solution can be viewed on the course web-page.

3.2 Mean-Field Dynamo Models

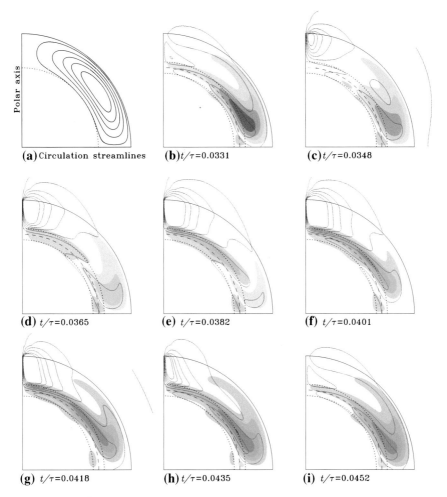

Fig. 3.10 Snapshots (**a–i**) covering half a cycle of an $\alpha\Omega$ dynamo solution including meridional circulation, starting at the time of polarity reversal in the polar surface field. Meridional circulation streamlines are plotted on panel **a**, the flow being poleward at the surface and equatorward at the core–envelope interface. Color coding of the toroidal field and poloidal fieldlines as on Fig. 3.7. This α-quenched solution uses the same differential rotation, diffusivity, and α-effect profiles as on Fig. 3.7, with parameter values $C_\alpha = 0.5$, $C_\Omega = 5 \times 10^5$, $\Delta\eta = 0.1$, $R_m = 2500$. Note the strong amplification of the surface polar fields, the latitudinal stretching of poloidal fieldlines by the meridional flow at the core–envelope interface, and the weak, secondary dynamo mode in the equatorial region of the tachocline.

around $R_m \sim 10^2$ that persists at least up to $R_m = 5000$; the solution of Fig. 3.9c, develops a dominant equatorial branch at $R_m \sim 200$, but a dominant high-latitude branch takes over from $R_m \sim 10^3$ onward.

Figure 3.10b through i shows half a cycle of our $\alpha \propto \cos\theta$ reference solution, now for parameter values $C_\alpha = 0.5$, $C_\Omega = 5 \times 10^5$, $\Delta\eta = 0.1$, and $R_m = 2500$, which for an envelope diffusivity reduced to $\eta_e = 5 \times 10^6$ m^2 s^{-1} corresponds to a solar-like surface poleward flow and differential rotation. The transport of the magnetic field by meridional circulation is clearly apparent, and concentrates the toroidal field to low latitudes, which is great from the point of view of the sunspot butterfly diagram. Note also how poloidal fieldlines suffer very strong stretching in the latitudinal direction within the tachocline (panels c through f), a direct consequence of shearing—in addition to plain transport—by the equatorward flow. One interesting consequence is that induction of the toroidal field is now effected primarily by the *latitudinal* shear within the tachocline, with the radial shear, although larger in magnitude, playing a lesser role since $B_r/B_\theta \ll 1$.

The meridional flow also has a profound impact on the magnetic field evolution at $r = R$, as it concentrates the poloidal field in the polar regions. This leads to a large amplification factor through magnetic flux conservation, so that dynamo solutions such as shown on Fig. 3.10 are typically characterized by very large polar field strengths, here 0.07 T, for an equipartition field strength $B_{eq} = 0.5$ T in Eq. (3.50). This is only a factor of 4 or so smaller than the toroidal field in the tachocline, even though we have here $C_\alpha/C_\Omega = 10^{-6}$. This concentrated poloidal field, when advected downwards to the polar regions of the tachocline, is responsible for the strong polar branch often seen in the butterfly diagram of dynamo solutions including a rapid meridional flow.

It is noteworthy that to produce a butterfly-like time-latitude diagram of the toroidal field at the core–envelope interface, the required value of R_m in conjunction with the observed surface meridional flow speed and reasonable profile for the internal return flow, ends up requiring a rather low envelope magnetic diffusivity, $\lesssim 10^7$ m^2 s^{-1}, which stands at the very low end of the range suggested by mean-field estimates such as provided by Eq. (3.22). Still, kinematic $\alpha\Omega$ mean-field models including meridional circulation and simple algebraic α-quenching can produce equatorially-concentrated and equatorially propagating dynamo modes with a period resembling that of the solar cycle for realistic, solar-like differential rotation and circulation profiles. Nice and fine, but it turns out we have another potential problem on our hands.

3.2.12 Interface Dynamos

The α-quenching expression (Eq. 3.50) used in the two preceding sections amounts to saying that dynamo action saturates once the mean, dynamo-generated field reaches an energy density comparable to that of the driving turbulent fluid motions, i.e., $B_{eq} \sim \sqrt{\mu_0 \varrho} u_t$, where u_t is the turbulent velocity amplitude. This appears eminently sensible, since from that point on a toroidal fieldline would have sufficient tension to resist deformation by cyclonic turbulence, and so could no longer feed the α-effect. At the base of the solar convective envelope, one finds $B_{eq} \sim 1$ T, for $u_t \sim 10$ m s^{-1}, according to standard mixing length theory of convection. However,

3.2 Mean-Field Dynamo Models

various calculations and numerical simulations have indicated that long before the mean toroidal field B reaches this strength, the helical turbulence reaches equipartition with the *small-scale*, turbulent component of the magnetic field. Such calculations also indicate that the ratio between the small-scale and mean magnetic components should itself scale as $R_m^{1/2}$, where $R_m = u_t\ell/\eta$ is a magnetic Reynolds number based on the turbulent speed u_t but *microscopic* magnetic diffusivity. This then leads to the alternate quenching expression

$$\alpha \to \alpha(B) = \frac{\alpha_0}{1 + R_m(B/B_{eq})^2}, \tag{3.51}$$

known in the literature as *strong α-quenching* or *catastrophic quenching*. Since $R_m \sim 10^8$ in the solar convection zone, this leads to quenching of the α-effect for very low amplitudes of the mean magnetic field, of order 10^{-5} T. Even though significant field amplification is likely in the formation of a toroidal flux rope from the dynamo-generated magnetic field, we are now a very long way from the 6–16 T demanded by simulations of buoyantly rising flux ropes and sunspot formation.

A way out of this difficulty was proposed by E.N. Parker in the form of *interface dynamos*. The idea is beautifully simple: if the toroidal field quenches the α-effect, amplify and store the toroidal field away from where the α-effect is operating! Parker showed that in a situation where a radial shear and the α-effect are segregated on either side of a discontinuity in magnetic diffusivity taken to coincide with the core–envelope interface, the constant coefficient $\alpha\Omega$ dynamo equations considered already in Sect. 3.2.6 support solutions in the form of traveling surface waves localized on the discontinuity. The key aspect of Parker's (linear, cartesian, analytical) solution is that for supercritical dynamo waves, the ratio of peak toroidal field strength on either side of the discontinuity surface is found to scale with the diffusivity ratio as

$$\frac{\max(B_e)}{\max(B_c)} \sim \left(\frac{\eta_e}{\eta_c}\right)^{-\frac{1}{2}}. \tag{3.52}$$

If the core diffusivity η_c assumes the microscopic value, and the envelope diffusivity (η_e) is of turbulent origin so that $\eta_e \sim \ell u_t$, then the toroidal field strength ratio scales as $\sim (u_t\ell/\eta_c)^{1/2} \equiv R_m^{1/2}$. This is precisely the factor needed to bypass strong α-quenching, at least as embodied in Eq. (3.51).

As an illustrative example, Fig. 3.11a shows a series of radial cuts of the toroidal magnetic component at 15° latitude, spanning half a cycle in a numerical interface solution with $C_\Omega = 2.5 \times 10^5$, $C_\alpha = +10$, and a core-to-envelope diffusivity contrast $\Delta\eta = 10^{-2}$. The differential rotation and magnetic diffusivity profiles are the same as before, but here the α-effect is now (even more artificially) concentrated towards the equator, by imposing a latitudinal dependency $\alpha \sim \sin(4\theta)$ for $\pi/4 \leq \theta \leq 3\pi/4$, and zero otherwise.

This model does achieve the kind of toroidal field amplification one would like to see in interface dynamos. Notice how the toroidal field peaks below the core–envelope interface (vertical dotted line), well below the α-effect region and near the

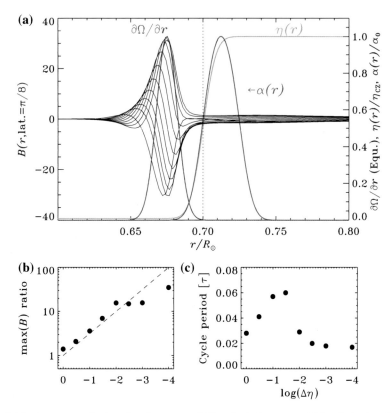

Fig. 3.11 A representative interface dynamo model in spherical geometry. This solution has $C_\Omega = 2.5 \times 10^5$, $C_\alpha = +10$, and a core-to-envelope diffusivity contrast of 10^{-2}. Panel **a** shows a series of radial cuts of the toroidal field at latitude 15°. The (normalized) radial profiles of magnetic diffusivity, α-effect, and radial shear are also shown, again at latitude 15°. The core–envelope interface is again at $r/R = 0.7$ (*dotted line*), where the magnetic diffusivity varies near-discontinuously. Panels **b** and **c** show the variations of the core-to-envelope peak toroidal field strength and dynamo period with the diffusivity contrast, for a sequence of otherwise identical dynamo solutions. Adapted from Living Review of Charbonneau (2010).

peak in radial shear. Figure 3.11b shows how the ratio of peak toroidal field below and above r_c varies with the imposed diffusivity contrast $\Delta\eta$. The dashed line is the dependency expected from Eq. (3.52). For a relatively low diffusivity contrast, $-1.5 \lesssim \log(\Delta\eta) \lesssim 0$, both the toroidal field ratio and dynamo period increase as $\sim (\Delta\eta)^{-1/2}$. Below $\log(\Delta\eta) \sim -1.5$, the max($B$)-ratio increases more slowly, and the cycle period falls, as can be seen on Fig. 3.11c. This is basically an electromagnetic skin-depth effect; unlike in the original picture proposed by Parker, here the poloidal field must diffuse down a finite distance into the tachocline before shearing into a toroidal component can commence. With this distance set by our adopted profile of $\Omega(r,\theta)$, as $\Delta\eta$ becomes very small there comes a point where the dynamo period is such that the poloidal field cannot diffuse as deep as the peak in radial shear in

3.2 Mean-Field Dynamo Models 121

the course of a half cycle. The dynamo then runs on a weaker shear, thus yielding a smaller field strength ratio and weaker overall cycle.

3.3 Babcock–Leighton Models

Solar cycle models based on what is now called the Babcock–Leighton mechanism were first developed in the early 1960s, yet they were temporarily eclipsed by the rise of mean-field electrodynamics a few years later. Their revival was motivated in part by the fact that synoptic magnetographic monitoring over solar cycles 21 and 22 has offered strong evidence that the surface polar field reversals are triggered by the decay of active regions (see Fig. 3.4). The crucial question is whether this is a mere side-effect of dynamo action taking place independently somewhere in the solar interior, or a dominant contribution to the dynamo process itself.

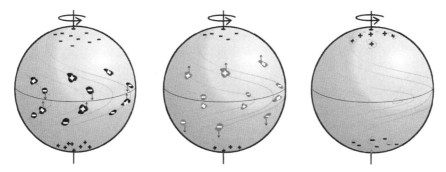

Fig. 3.12 Cartoon of the Babcock–Leighton mechanism. At *left*, a number of bipolar magnetic regions (BMR) have emerged, with opposite leading/following polarity patterns in each hemisphere, as per Hale's polarity law. After some time (*middle*), the BMRs have started decaying, with the leading components experiencing diffusive cancellation across the equator, while the trailing components have moved to higher latitudes. At later time, (*right*), the net effect is the buildup of an hemispheric flux of opposite polarity in the N and S hemisphere, i.e., a net dipole moment (see text). Diagram kindly provided by D. Passos.

Figure 3.12 illustrates the basic idea of the Babcock–Leighton mechanism. Consider the bipolar magnetic regions (BMR) sketched on the left. Recall that each of these is the photospheric manifestation of a toroidal flux rope emerging as an Ω-loop (see Fig. 3.3). The *leading* (*trailing*) component of each BMR is that located ahead (behind) with respect to the direction of the sun's rotation. Joy's law states that, on average, the leading component is located at lower latitude than the trailing component, so that a line joining each component of the pair makes an angle with respect to the E–W line. Hale's polarity law also informs us that the leading/trailing magnetic polarity pattern is opposite in each hemisphere, a reflection of the equatorial antisymmetry of the underlying toroidal flux system. Horace W. Babcock (1912–2003) demonstrated empirically from his early magnetographic observation of the sun's surface magnetic field that as the BMRs decay (presumably under the influence of turbulent convection), the trailing components drift to higher latitudes, leaving the leading components at lower latitudes, as sketched on Fig. 3.12 (middle). Babcock also argued that the trailing polarity flux released to high latitude by the cumulative effects of the emergence and subsequent decay of many BMRs was responsible for the reversal of the sun's large-scale dipolar field (right).

More germane from the dynamo point of view, the Babcock–Leighton mechanism taps into the (formerly) toroidal flux in the BMRs to produce a poloidal magnetic component. To the degree that a positive dipole moment is being produced from a toroidal field that is positive in the N-hemisphere, this is a bit like a positive α-effect in mean-field theory. In both cases the Coriolis force is the agent imparting a twist on a magnetic field; with the α-effect this process occurs on the small spatial scales and operates on individual magnetic fieldlines. In contrast, the Babcock–Leighton mechanism operates on the large scales, the twist being imparted via the the Coriolis force acting on the flow generated along the axis of a buoyantly rising magnetic flux tube.

3.3.1 Sunspot Decay and the Babcock–Leighton Mechanism

Evidently this mechanism can operate as sketched on Fig. 3.12 provided the magnetic flux in the leading and trailing components of each (decaying) BMR are separated in latitude faster than they can diffusively cancel with one another. Moreover, the leading components must end up at low enough latitudes for diffusive cancellation to take place across the equator. This is not trivial to achieve, and we now take a more quantitative look at the Babcock–Leighton mechanism, first with a simple 2D numerical model.

The starting point of the model is the grand sweeping assumption that, once the sunspots making up the bipolar active region lose their cohesiveness, their subsequent evolution can be approximated by the passive advection and resistive decay of the radial magnetic field component. This drastic simplification does away with any dynamical effect associated with magnetic tension and pressure within the spots, as well as any anchoring with the underlying toroidal flux system. The model is further simplified by treating the evolution of B_r as a two-dimensional transport problem on a spherical surface corresponding to the solar photosphere. Consequently, no subduction of the radial field can take place.

Even under these simplifying assumptions, the evolution is still governed by the MHD induction equation, specifically its r-component. The imposed flow is made of an axisymmetric surface "meridional circulation", basically a poleward-converging flow in the latitudinal direction on the sphere, and differential rotation in the azimuthal direction:

$$\boldsymbol{u}(\theta) = u_\theta(\theta)\hat{\boldsymbol{e}}_\theta + \Omega_S(\theta) R \sin\theta \hat{\boldsymbol{e}}_\phi , \qquad (3.53)$$

where Ω_S is the solar-like surface differential rotation profile used in the preceding chapter (see Eq. 2.28). Note that in general $\nabla \cdot \boldsymbol{u} \neq 0$ here, a direct consequence of

3.3 Babcock–Leighton Models

working on a spherical surface without possibility of subduction. The r-component of the induction equation is cast in non-dimensional form, by expressing length in units of the solar radius R, and time in units of $\tau_c = R/u_0$, i.e., the advection time associated with the meridional flow. Introducing a new latitudinal variable $\mu = \cos\theta$, neglecting all radial derivatives, and evaluating the resulting expression at $r/R = 1$ results in[9]:

$$\frac{\partial B_r}{\partial t} = \frac{\partial}{\partial \mu}\left[(1-\mu^2)^{1/2} u_\theta B_r\right] - \frac{\Omega_S}{R_u}\frac{\partial B_r}{\partial \phi}$$
$$+ \frac{1}{R_m}\left[\frac{\partial}{\partial \mu}\left((1-\mu^2)\frac{\partial B_r}{\partial \mu}\right) + \frac{1}{(1-\mu^2)}\frac{\partial^2 B_r}{\partial \phi^2}\right]. \qquad (3.54)$$

The solutions are defined in terms of the two nondimensional numbers:

$$R_m = \frac{u_0 R}{\eta}, \qquad R_u = \frac{u_0}{\Omega_0 R}, \qquad (3.55)$$

with u_0 a characteristic speed for the meridional flow, and η the net magnetic diffusivity, assumed constant over the spherical surface defining the solution domain. Using $\Omega_0 = 3 \times 10^{-6}$ rad s^{-1}, $u_0 = 15$ m s^{-1}, and $\eta = 6 \times 10^8$ m^{2}s^{-1} yields $\tau_c \simeq 1.5$ yr, $R_m \simeq 20$ and $R_u \simeq 10^{-2}$. The former is really a measure of the (turbulent) magnetic diffusivity, and is the only free parameter of the model, as u_0 is well constrained by surface Doppler measurements. The corresponding magnetic diffusion time is $\tau_\eta = R^2/\eta \simeq 26$ yr, so that $\tau_c/\tau_\eta \ll 1$.

Figure 3.13 shows a representative solution,[10] computed assuming a simple analytic form for the meridional flow, namely $u_\theta(\theta) = 2u_0 \sin\theta \cos\theta$. The initial condition (panel a, $t = 0$) describes a series of eight BMRs, four per hemisphere, equally spaced 90° apart at latitudes $\pm 15°$. Each BMR consists of two Gaussian profiles of opposite sign and adding up to zero net flux, with angular separation $d = 10°$ and with a line joining the center of the two Gaussians tilted with respect to the E–W direction[11] by an angle γ, itself related to the latitude θ_0 of the BMR's midpoint according to the Joy law-like relation:

$$\sin\gamma = 0.5\cos\theta_0. \qquad (3.56)$$

---

[9] This expression is best obtained through the original form of the MHD induction equation (1.59), rather than the (usually) equivalent form where the $-\nabla \times (\eta\nabla \times \mathbf{B})$ is expressed as $\eta\nabla^2 \mathbf{B}$; the distinction hinges on the metric terms resulting from the application of the Laplacian operator on a vector quantity. Some have to be forced to zero if radial diffusion is explicitly ignored. More specifically, in the case considered here this would mean zeroing the $-2B_r/r^2$ term in the diffusive term on the RHS of the r-component of the induction equation in spherical coordinates, as given in Appendix B.

[10] An animation of which can of course be viewed on the course web-page.

[11] Remember that this is meant to represent the result of a toroidal flux rope erupting through the surface, so that in this case the underlying toroidal field is positive in the Northern hemisphere, which is the polarity of the trailing "spot", as measured with respect to the direction of rotation, from left to right here.

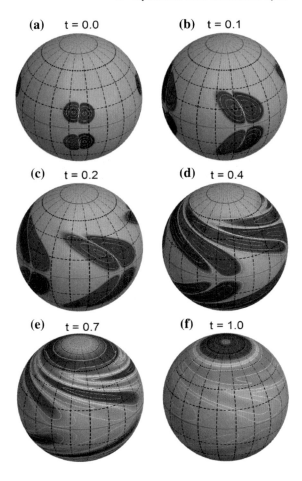

Fig. 3.13 Evolution (**a**–**f**) of the surface radial magnetic field for two sets of four BMRs equally spaced in longitude, and initially located at latitudes ±15°, with opposite polarity ordering in each hemisphere, as per Hale's polarity laws. The surface field evolves in response to diffusion and advective transport by differential rotation and a poleward meridional flow, as described by the 2D advection-diffusion equation (3.54). Parameter values are $R_u = 10^{-2}$ and $R_m = 50$, with time given in units of the meridional flow's characteristic time $\tau_c = R/u_0$.

The symmetry of the flow and initial condition on $B_r(\theta, \phi)$ means that the problem can be solved in a single hemisphere with $B_r = 0$ enforced in the equatorial plane, in a 90° wide longitudinal wedge with periodic boundary conditions in ϕ.

The combined effect of circulation, diffusion and differential rotation is to concentrate the magnetic polarity of the trailing "spot" to high latitude. The polarity of the leading spot dominates at lower latitudes, but experiences diffusive cancellation with the opposite polarity leading flux from its "cousin" in the other solar hemisphere. It is this cross-equatorial diffusive cancellation that is ultimately responsible for the buildup of a net hemispheric flux. At mid-latitudes, the effect of differential rotation is to stretch longitudinally the unipolar regions originally associated with each member of the BMR, causing the development of thin banded structures of opposite magnetic polarities. This leads to thus enhanced dissipation, much like in the cellular flow problem considered earlier in Sect. 2.3.

3.3 Babcock–Leighton Models

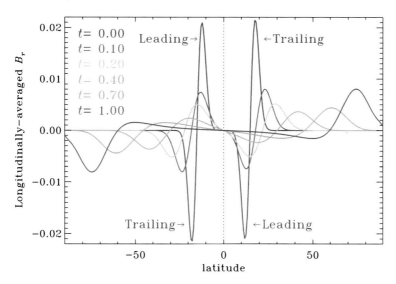

Fig. 3.14 Latitudinal profile of the longitudinally averaged vertical magnetic field, at the six epochs plotted on Fig. 3.13. The strong signal at $t = 0$ results entirely from the slight misalignment of the emerging BMRs with respect to the E–W direction. By one turnover time, two polar caps of oppositely-signed magnetic field have built up, amounting to a net dipole moment (see text).

The combined effects of these advection-diffusion processes is to separate in latitude the two polarities of the BMR. This is readily seen upon calculating the longitudinally averaged latitudinal profiles of B_r, as shown on Fig. 3.14 for the same six successive epochs corresponding to the snapshots on Fig. 3.13. The poleward displacement of the trailing polarity "bump" is the equivalent to Babcock's original cartoon (cf. Fig. 3.12). The time required to achieve this here is $t/\tau_c \sim 1$, and scales as $(R_m/R_u)^{1/3}$. The significant amplification of the trailing polarity bump from $t/\tau_c \gtrsim 0.5$ onward is a direct consequence of magnetic flux conservation in the poleward-converging meridional flow. Notice also the strong latitudinal gradient in B_r at the equator (dotted line) early in the evolution; the associated trans-equatorial diffusive polarity cancellation affects preferentially the leading spots of each pairs, since the trailing spots are located slightly farther away from the equator.

Consider again the mean signed and unsigned magnetic flux:

$$\Phi = |\langle B_r \rangle|, \qquad F = \langle |B_r| \rangle, \tag{3.57}$$

where the averaging operator is now defined on the spherical surface, for the Northern and Southern hemispheres separately:

$$\langle B_r \rangle = \int_0^{2\pi} \int_{-\pi/2(0)}^{0(\pi/2)} B_r(\theta, \phi) \sin\theta \, d\theta \, d\phi. \tag{3.58}$$

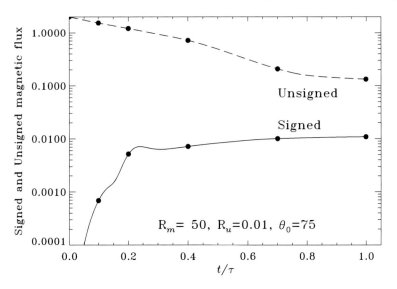

Fig. 3.15 Evolution of the Northern hemisphere signed (*solid line*) and unsigned (*dashed line*) magnetic flux for the solution of Fig. 3.13. The *solid dots mark* the times at which the snapshots and longitudinal averages are plotted on Figs. 3.13 and 3.14.

Figure 3.15 shows the time-evolution of the signed (Φ, solid line) and unsigned (F, dashed) fluxes in the Northern hemisphere, for the solution of Fig. 3.13. The unsigned flux decreases rapidly at first, then settles into a slower decay phase. Meanwhile a small but significant hemispheric signed flux is building up. This is a direct consequence of (negative) flux cancellation across the equator, mediated by diffusion, and is the Babcock–Leighton mechanism in action. Note the dual, conflicting role of diffusion here; it is needed for cross-hemispheric flux cancellation, yet must be small enough to allow the survival of a significant trailing polarity flux on timescales of order τ_c.

The efficiency (Ξ) of the Babcock–Leighton mechanism, i.e., converting toroidal to poloidal field, can be defined as the ratio of the signed flux at $t = \tau_c$ to the BMR's initial unsigned flux:

$$\Xi = 2 \frac{\Phi(t = \tau_c)}{F(t = 0)} . \tag{3.59}$$

Note that Ξ is independent of the assumed initial field strength of the BMRs since Eq. (3.54) is linear in B_r. Looking back at Fig. 3.15, one would eyeball the efficiency at about 1% in converting the BMR flux to polar cap signed flux. This conversion efficiency turns out to be a rather complex function of BMR parameters; it is expected to grow as with increasing tilt γ, and therefore should increase with latitudes as per Joy's law, yet proximity to the equator favors trans-equatorial diffusive flux cancellation of the leading component; moreover, having $du_\theta/d\theta < 0$ favors the separation of the two BMR components, thus minimizing diffusive flux cancellation between the leading and trailing components. These competing effects lead to a toroidal-to-poloidal conversion efficiency peaking for BMRs emerging at fairly low latitudes, the exact value depending on the latitudinal variation of the adopted surface meridional flow profile. At any rate, we noted already (Sect. 3.1) that the sun's polar cap flux peaks at solar minimum, at a value amounting to $\sim 0.1\%$ of the cycle-integrated active region (unsigned) flux; the efficiency required of the Babcock–Leighton mechanism is indeed quite modest.

3.3.2 Axisymmetrization Revisited

Take another look at Fig. 3.13; at $t = 0$ (panel a) the surface magnetic field distribution is highly non-axisymmetric. By $t/\tau_c = 0.7$ (panel e), however, the field distribution shows a far less pronounced ϕ-dependency, especially at high latitudes where in fact B_r is nearly axisymmetric, and by $t/\tau_c = 1$ (panel f) there is little non-axisymmetric field left over the surface. This should remind you of something we encountered earlier: axisymmetrization of a non-axisymmetric magnetic field by an axisymmetric differential rotation (Sect. 2.3.5), the spherical analog of flux expulsion. In fact a closer look at the behavior of the unsigned flux on Fig. 3.15 (dashed line) already shows a hint of the two-timescale behavior we have come to expect of axisymmetrization: the rapid destruction of the non-axisymmetric flux component and slower ($\sim \tau_\eta$) diffusive decay of the remaining axisymmetric flux distribution.

Since the spherical harmonics represent a complete and nicely orthonormal functional basis on the sphere, it follows that the initial condition for the simulation of Fig. 3.13 can be written as

$$B_r^0(\theta, \phi) = \sum_{l=0}^{\infty} \sum_{m=-l}^{+l} b_{lm} Y_{lm}(\theta, \phi), \qquad (3.60)$$

where the Y_{lm}'s are the spherical harmonics:

$$Y_{lm}(\theta, \phi) = \sqrt{\frac{2l+1}{4\pi} \frac{(l-m)!}{(l+m)!}} P_l^m(\cos\theta) e^{im\phi}, \qquad (3.61)$$

and with the coefficients b_{lm} given by

$$b_{lm} = \int_0^{2\pi} \int_0^{\pi} B_r^0(r, \theta) Y_{lm}^*(\theta, \phi) \, d\theta \, d\phi, \qquad (3.62)$$

where the "$*$" indicates complex conjugation. Now, axisymmetrization will wipe out all $m \neq 0$ modes, leaving only the $m = 0$ modes to decay away on the slower diffusive timescale.[12] Therefore, at the end of the axisymmetrization process, the radial field distribution now has the form:

$$B_r(\theta) = \sum_{l=0}^{\infty} \sqrt{\frac{2l+1}{4\pi}}\, b_{l0} P_l^0(\cos\theta)\,, \qquad t/\tau_c \gg R_u\,, \qquad (3.63)$$

which now describes an axisymmetric poloidal magnetic field.[13] Voilà!

3.3.3 Dynamo Models Based on the Babcock–Leighton Mechanism

So now we understand how the Babcock–Leighton mechanism can convert a toroidal magnetic field into a poloidal component, and therefore act as a poloidal source term in Eq. (2.61). Now we need to construct a solar cycle model based on this idea. One big difference with the $\alpha\Omega$ models considered in Sect. 3.2 is that the two source regions are now spatially segregated: production of the toroidal field takes place in the tachocline, as before, but now production of the poloidal field takes place in the surface layers.

The mode of operation of a generic solar cycle model based on the Babcock–Leighton mechanism is illustrated in cartoon form on Fig. 3.16. Let P_n represent the amplitude of the high-latitude, surface ("A") poloidal magnetic field in the late phases of cycle n, i.e., after the polar field has reversed. The poloidal field P_n is advected downward by meridional circulation (A→B), where it then starts to be sheared by the differential rotation while being also advected equatorward (B→C). This leads to the growth of a new low-latitude (C) toroidal flux system, T_{n+1}, which becomes buoyantly unstable (C→D) and starts producing sunspots (D), which subsequently decay and release the poloidal flux P_{n+1} associated with the new cycle $n+1$. Poleward advection and accumulation of this new flux at high latitudes (D→A) then obliterates the old poloidal flux P_n, and the above sequence of steps begins anew. Meridional circulation clearly plays a key role in this "conveyor belt" model of the solar cycle, by providing the needed link between the two spatially segregated source regions. Under this configuration, Babcock–Leighton solar cycle models operate as flux-transport dynamos.

---

[12] With $u = 0$, the decay rate of those remaining modes are given by the eigenvalues of the 2D pure resistive decay problem, much like in Sect. 2.1.

[13] Note however that the above expression does not include the advective effect of the meridional flow, so that the axisymmetric distribution on Fig. 3.13f is more sharply peaked at high latitudes than what one would infer from Eq. (3.63) applied to the initial radial field profile plotted on Fig. 3.13a.

3.3 Babcock–Leighton Models

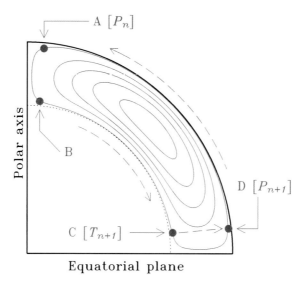

Fig. 3.16 Operation of a solar cycle model based on the Babcock–Leighton mechanism. The diagram is drawn in a meridional quadrant of the sun, with streamlines of meridional circulation plotted in blue. Poloidal field having accumulated in the surface polar regions ("A") at cycle n must first be advected down to the core–envelope interface (*dotted line*) before production of the toroidal field for cycle $n + 1$ can take place (B→C). Buoyant rise of flux rope to the surface (C→D) is a process taking place on a much shorter timescale. From Living Review of Charbonneau (2010).

3.3.4 The Babcock–Leighton Poloidal Source Term

The definition of the Babcock–Leighton source term S to be inserted in Eq. (2.61) is evidently the crux of the model. The dynamo solutions presented in what follows use the following:

$$S(r, \theta, B(t)) = s_0 f(r) g(\theta) h(B) B(r_c, \theta, t) \tag{3.64}$$

with

$$f(r) = \frac{1}{2}\left[1 + \mathrm{erf}\left(\frac{r - r_2}{d_2}\right)\right]\left[1 - \mathrm{erf}\left(\frac{r - r_3}{d_3}\right)\right], \tag{3.65}$$

$$g(\theta) = \sin\theta \cos\theta, \tag{3.66}$$

$$h(B) = \left[1 + \mathrm{erf}\left(\frac{B(r_c, \theta, t) - B_1}{w_1}\right)\right]\left[1 - \mathrm{erf}\left(\frac{B(r_c, \theta, t) - B_2}{w_2}\right)\right], \tag{3.67}$$

where s_0 is a numerical coefficient setting the strength of the source term (corresponding dynamo number being $C_S = s_0 R/\eta_e$), and with the various remaining numerical coefficients taking the values $r_2/R = 0.95$, $r_3/R = 1$, $d_2/R = d_3/R = 10^{-2}$, $B_1 = 6$, $B_2 = 10$, $w_1 = 2$, and $w_2 = 8$, the latter four all measured in tesla. Note that the dependency on B is *non-local*, i.e., it involves the toroidal field evaluated at the core–envelope interface r_c, (but at the same polar angle θ). This nonlocality in B represents the fact that the strength of the source term is proportional to the field strength in the bipolar active region, itself presumably reflecting the strength of the diffuse toroidal field near the core–envelope interface, where the magnetic flux ropes eventually giving rise to the bipolar active region originate. The combination of error functions in Eq. (3.67) restricts the operating range of the model to a finite interval in toroidal field strength, and is motivated by simulations of the stability and buoyant rise of thin flux tubes, as discussed in Sect. 3.1.4. Other equally reasonable prescriptions and modelling approaches are of course possible (see bibliography at the end of this chapter).

At any rate, inserting this source term into Eq. (2.61) is what we need to bypass Cowling's theorem and produce a viable dynamo model. The nonlocality of S notwithstanding, at this point the model equations are definitely mean-field like. Yet no averaging on small scales is involved. What is implicit in Eq. (3.64) is some sort of averaging process at least in longitude and time, over many BMR emergences.

3.3.5 A Sample Solution

Figure 3.17 shows a series of meridional quadrant snapshot of one such Babcock–Leighton dynamo solution, in the now usual format.[14] The figure covers a half-cycle, corresponding to one sunspot cycle, starting approximately at the time one would identify as sunspot minimum, with sunspot maximum (based on magnetic energy as a proxy for the sunspot number) occurring between panels (e) and (f), and reversal of the polar field shortly thereafter, between panels (f) and (g). As with the advection-dominated $\alpha\Omega$ solution of the preceding section, this solution is characterized by an equatorward propagation of the toroidal field in the tachocline driven by the meridional flow. The turnover time of the meridional flow is here again the primary determinant of the cycle period. With $\eta_e = 3 \times 10^7 \, \text{m}^2 \, \text{s}^{-1}$, this solution has a nicely solar like half period of 12.4 yr. All in all, this is once again a reasonable representation of the cyclic spatiotemporal evolution of the solar large-scale magnetic field.

The strong toroidal fields building up within the polar regions of the tachocline in the course of the cycle (see panel c through g on Fig. 3.17) are entirely unrelated to the adopted latitudinal dependency of the Babcock–Leighton source term. It results instead from the strong polar field advected downwards by the meridional flow, inducing a toroidal component through the inductive action of both the latitudinal shear within the convective envelope, and the negative radial shear in the polar regions of the tachocline. Here this toroidal component mostly decays away under the influence of Ohmic dissipation, and contributes very little to the production of

---

[14] ...and guess where you have to go to view an animation of this solution...?

3.3 Babcock–Leighton Models

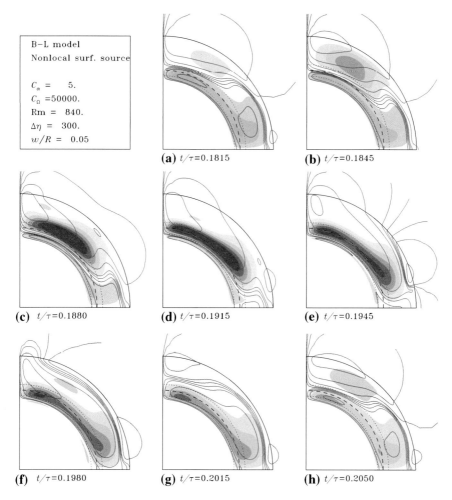

Fig. 3.17 Snapshots (**a–h**) covering half a cycle of a Babcock–Leighton dynamo solution, as described in the text. Color coding of the toroidal field and poloidal fieldlines as on Fig. 3.7. This solution uses the same differential rotation, magnetic diffusivity and meridional circulation profile as for the advection-dominated $\alpha\Omega$ solution of Sect. 3.2.11, but now with the non-local surface source term defined through Eq. (3.64), with parameter values $C_\alpha = 5$, $C_\Omega = 5 \times 10^4$, $\Delta\eta = 0.003$, $R_m = 840$. Note again the strong amplification of the surface polar fields, the latitudinal stretching of poloidal fieldlines by the meridional flow at the core–envelope interface.

the next cycle's poloidal component, which builds up at lower latitude (panel e) and is then carried poleward by the meridional flow (panels e→h).

Although it exhibits the desired equatorward propagation, the toroidal field butterfly diagram on Fig. 3.17 peaks at much higher latitude ($\sim 45°$) than the sunspot butterfly diagram ($\sim 15°$–$20°$). This occurs because this is a solution with high magnetic diffusivity contrast, where meridional circulation closes at the core–envelope interface, so that the *latitudinal* component of differential rotation dominates the production of the toroidal field. This difficulty can be alleviated by letting the meridional circulation penetrate below the core–envelope interface, but this often leads to the production of a strong polar branch, again a consequence of both the strong radial shear present in the high-latitude portion of the tachocline, and of the concentration of the poloidal field taking place in the high latitude-surface layer prior to this field being advected down into the tachocline by meridional circulation (viz. Figs. 3.16 and 3.17). Another interesting option to avoid excessive polar field amplification is to rely on turbulent pumping to carry the surface field downward into the convection zone faster than it can accumulate at the poles.

A noteworthy property of this class of model is the dependency of the cycle period on model parameters; over a wide portion of parameter space, the meridional flow speed is found to be the primary determinant of the cycle period (P). This behavior arises because, in these models, the two source regions are spatially segregated, and the time required for circulation to carry the poloidal field generated at the surface down to the tachocline is what effectively sets the cycle period. The corresponding time delay introduced in the dynamo process has rich dynamical consequences, to be discussed in Sect. 4.4 below. On the other hand, P is found to depend very weakly on the assumed values of the source term amplitude s_0, and turbulent diffusivity η_e; this is very much unlike the behavior typically found in mean-field models, where P scales nearly as η_e^{-1} in α-quenched $\alpha\Omega$ mean-field models.

3.4 Models Based on HD and MHD Instabilities

In the presence of stratification and rotation, a number of hydrodynamical (HD) and magnetohydrodynamical instabilities associated with the presence of a strong toroidal field in the stably stratified, radiative portion of the tachocline can lead to the growth of disturbances with a net kinetic helicity. Under suitable circumstances, such disturbances can act upon a pre-existing large-scale magnetic field component to produce a toroidal electromotive force, and therefore act as a source of poloidal field.

Different types of solar cycle models have been constructed in this manner, two promising ones being briefly reviewed in this section. In both cases the resulting dynamo models end up being described by something closely resembling our now well-known axisymmetric mean-field dynamo equations, the novel poloidal field regeneration mechanisms being once again subsumed in an α-effect-like source term appearing of the RHS of Eq. (2.61).

3.4.1 Models Based on Shear Instabilities

Hydrodynamical stability analyses of the latitudinal shear profile in the solar tachocline indicate that the latter may be unstable to non-axisymmetric perturbations, with the instabilities planforms characterized by a net kinetic helicity.

3.4 Models Based on HD and MHD Instabilities

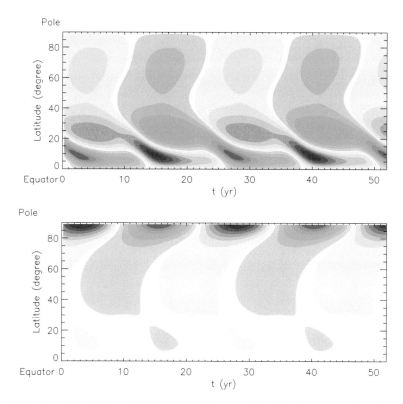

Fig. 3.18 Time–latitude "butterfly" diagrams of the toroidal field at the core–envelope interface (*top*), and surface radial field (*bottom*) for a representative dynamo solution with the tachocline α-effect of Dikpati and Gilman (2001). This solution has a solar-like half-period of 11 yr. Note how the deep toroidal field peaks at very low latitudes, in good agreement with the sunspot butterfly diagram. For this solution the equatorial deep toroidal field and polar surface radial field lag each other by $\sim \pi$, but other parameter settings can bring this lag closer to the observed $\pi/2$. Diagrams kindly provided by M. Dikpati.

Loosely inspired by Eq. (3.16), this allows the construction of an azimuthally-averaged α-effect-like source term that is directly proportional to the large-scale toroidal magnetic field component, just as in mean-field electrodynamics. The associated dynamo model is then described by the $\alpha\Omega$ form of the mean-field dynamo equations, including the meridional flow for the specific model considered here.

Figure 3.18 shows representative time-latitude diagrams of the toroidal magnetic field at the core–envelope interface, and surface radial field, for a flux transport dynamo solution based on this poloidal source mechanism. This is a solar-like solution with a mid-latitude surface meridional (poleward) flow speed of 17 m s^{-1}, envelope diffusivity $\eta_e = 5 \times 10^7$ m^2 s^{-1}, a core-to-envelope magnetic diffusivity contrast $\Delta\eta = 10^{-3}$, and a simple α-quenching-like amplitude-limiting nonlinearity.[15] Note the equatorward migration of the deep toroidal field, set here by the meridional flow in the deep envelope, and the poleward migration and intensification of the surface poloidal field, again a direct consequence of advection by meridional circulation, as in the mean-field dynamo models discussed in Sect. 3.2.11 in the advection-dominated, high R_m regime. The three-lobe structure of each spatiotemporal cycle in the butterfly diagram reflects the latitudinal structure in kinetic helicity profiles associated with the instability planforms.

The primary weakness of these models, in their present form, is their reliance on a linear stability analysis that altogether ignores the destabilizing effect of magnetic fields, especially since stability analyses have shown that the MHD version of the instability is easier to excite for toroidal field strengths of the magnitude believed to characterize the solar tachocline. Moreover, the planforms in the MHD version of the instability are highly dependent on the assumed underlying toroidal field profile, so that the kinetic helicity can be expected to (1) have a time-dependent latitudinal distribution, and (2) be intricately dependent on the mean toroidal field in a manner that is unlikely to be reproduced by a simple amplitude-limiting quenching formula.

3.4.2 Models Based on Flux-Tube Instabilities

As briefly discussed in Sect. 3.1, modelling of the rise of thin toroidal flux tubes (or ropes) throughout the solar convection zone has met with great success, in particular in reproducing the latitudes of emergence and tilt angles of bipolar sunspot pairs. It is also possible to use the thin-flux tube approximation to study the stability of toroidal flux ropes stored immediately below the base of the convection zone, to investigate the conditions under which they can actually be destabilized and give rise to sunspots. Once the tube destabilizes, calculations show that under the influence of rotation, the correlation between the flow and field perturbations is such as to yield a mean azimuthal electromotive force, equivalent to a *positive* α-effect in the N–hemisphere.

Figure 3.19 shows a stability diagram for this flux tube instability, in the form of growth rate contours in a 2D parameter space comprised of flux tube strength and latitudinal position at the core–envelope interface. The key is now to identify regions where weak instability arises (growth rates $\gtrsim 1$ yr). In the case shown on Fig. 3.19, these regions are restricted to flux tube strengths in the approximate range 6–15 T.

Although it has not yet been comprehensively studied, this dynamo mechanism has a number of very attractive properties. It operates without difficulty in the strong field regime (in fact in *requires* strong fields to operate). It also naturally yields dynamo action concentrated at low latitudes. Difficulties include the need of a relatively finely tuned magnetic diffusivity to achieve a solar-like dynamo period, and a relatively

---

[15] See the Dikpati and Gilman (2001) paper cited in the bibliography for more details.

3.4 Models Based on HD and MHD Instabilities

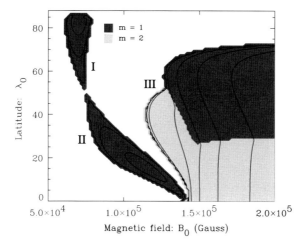

Fig. 3.19 Stability diagram for toroidal magnetic flux tubes located in the overshoot layer immediately beneath the core–envelope interface. The plot shows contours of growth rates in the latitude-field strength plane. The gray scale encodes the azimuthal wavenumber of the mode with largest growth rate, and regions left in *white* are stable. Dynamo action is associated with the regions with growth rates ~ 1 yr, here labeled I and II. Region III is associated with the rapid destabilisation, buoyant rise and emergence of magnetic flux, without significant dynamo action. Diagram kindly provided by A. Ferriz-Mas.

finely-tuned level of subadiabaticity in the overshoot layer for the instability to kick on and off at the appropriate toroidal field strengths.

The effects of meridional circulation in this class of dynamo models has yet to be investigated; this should be particularly interesting, since both analytic calculations and numerical simulations suggest a *positive* α-effect in the Northern-hemisphere, which should then produce *poleward* propagation of the dynamo wave at low latitude. Meridional circulation could then perhaps produce equatorward propagation of the dynamo magnetic field even with a positive α-effect, as it does in true mean-field models (cf. Sect. 3.2.11).

As an interesting aside, note on Fig. 3.19 how, except in a narrow range of field strengths around ~ 7 T, flux tubes located at high latitudes are always stable; this is due to the stabilizing effect of magnetic tension associated with high curvature of the toroidal flux ropes. Even if flux ropes were to form there, they may not necessarily show up at the surface as sunspots. This should be kept in mind when comparing time-latitudes diagrams produced by this or that dynamo model to the sunspot butterfly diagram; the two may not map onto one another as well as often implicitly assumed.

3.5 Global MHD Simulations

After this grand tour of (relatively) simple solar cycle models, it is worth briefly looking at the theoretical "real thing", namely global MHD simulations of thermally-driven convection in a thick, stratified rotating spherical shell, across which a solar

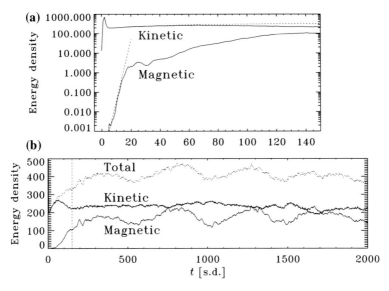

Fig. 3.20 Time series of kinetic and magnetic energy densities in a 3D anelastic MHD simulation of thermally-driven convection in a thick, stratified rotating spherical shell. Part **a** is a closeup on the first 150 solar days (one solar day ≡ 30 terrestrial days), and part **b** shows the evolution over 2,000 solar days (≡165 yr). The phase of exponential growth in the magnetic energy ($5 \lesssim t \lesssim 15$, slanted dotted line segment) begins once the convection has attained a statistically stationary state. The slower growth kicking in at $t \simeq 50$ is associated with the buildup of a large-scale magnetic component, which eventually develops cyclic polarity reversals, leading to the long-timescale modulation of the magnetic energy time series visible on part **b**.

heat flux is forced to flow. We focus in what follows on a specific set of simulations carried out in the anelastic regime.[16] The simulation domain includes most of the convection zone (here $0.718 \lesssim t/R \lesssim 0.96$), as well as a stably stratified fluid layer underneath ($0.61 \lesssim r/R \lesssim 0.718$). The background stratification is solar-like, and covers 4 scale heights in density, and radiation is treated in the diffusion approximation, as in Eq. (1.81).

Figure 3.20 shows time series of the kinetic and magnetic energies in a typical simulation, starting from a static, unmagnetized configuration ($\boldsymbol{u} = 0$, $\boldsymbol{B} = 0$) with small random seed magnetic field and velocity perturbation introduced at $t = 0$. Thermal convection sets in very rapidly, and leads to a rapid growth of kinetic energy across the convectively unstable part of the simulation domain in the first few solar days (1 sd ≡ 30 Earth days). Once convection has reached a statistically stationary state, small-scale dynamo action powered by this turbulent flow commences, and leads to the exponential growth of magnetic energy. Here, this phase of exponential growth lasts up to ~ 15 solar days, after which the Lorentz force starts to backreact

---

[16] The results presented here are taken pretty directly from the Ghizaru et al. (2010) and Racine et al. (2011) papers listed in the bibliography.

3.5 Global MHD Simulations

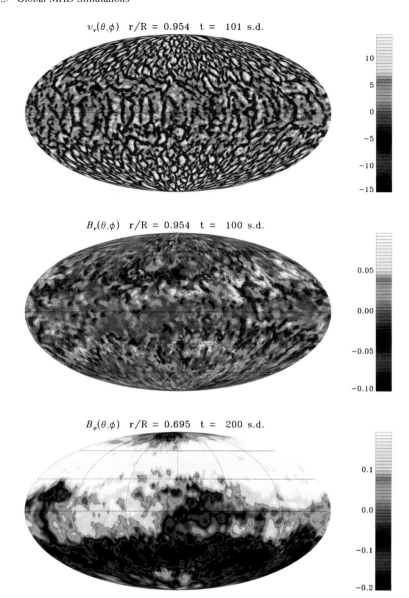

Fig. 3.21 Mollweide projections of the radial components of the flow velocity (*top*) and magnetic field (*middle*) on a spherical surface near the top of the simulation domain. Flow speeds are color-coded in m s^{-1}, and field strengths in tesla. Note the asymmetry in the upflow/downflow structures, and the relatively small spatial scale of the subsurface magnetic field. The *bottom* Mollweide projection shows the zonal magnetic component immediately beneath the base of the convective envelope, where a strong and well-organized axisymmetric component is already building up after 200 solar days. Adapted from Ghizaru et al. (2010) by permission of the AAS.

on the turbulent flow, leading to a saturation of the magnetic energy reminiscent of α-quenching (cf. Fig. 3.8), and completed at $\sim$ 20 solar days.

Figure 3.21 shows snapshots at $t = 100$ sd of the radial flow (top) and magnetic field (middle) components extracted near the top of the simulation domain. The morphological asymmetry between the broad, diffuse upflows and narrow concentrated downflows is quite typical of thermally-driven convection in a stratified environment. The magnetic field is swept horizontally in the broad areas of upflows and ends up preferentially concentrated in regions where downflow lanes meet, a feature that is typical of MHD convection. As with the CP flow solutions considered earlier, the subsurface magnetic field is spatially and temporally very intermittent, and is characterized by significant magnetic energy but very little net magnetic flux on large spatial scales.[17]

By this time, at least on the basis of these energy time series, one would judge the system to have reached a statistically stationary state. However, integrating further in time reveals variations setting in on longer timescales, associated with the slow buildup of a large-scale magnetic field carrying a net flux on those scales. This slower buildup is already apparent on Fig. 3.20. By about 100 solar days, the large-scale component has reached a strength such that it begins to quench the differential rotation having built up in the earlier phases of the simulation through the action of turbulent Reynolds stresses. This leads to a $\sim$ 20% drop in the kinetic energy density, from $\sim$ 270 to 230 J kg^{-1} at $t \simeq 150$ solar days.

This spatially well-organized magnetic component is particularly prominent at and beneath the base of the convective envelope, where the significant differential rotation, stably stratified environment, and injection of magnetic fields from above by downward turbulent pumping, all conspire to favor the buildup and accumulation of magnetic flux. The bottom Mollweide projection on Fig. 3.21 shows the zonal magnetic component at a depth slightly below the base of the convective envelope, at a later time in the simulation. Here the zonally-averaged toroidal field is seen to be well-organized on the larger scales, and in particular shows a clear antisymmetry with respect to the equatorial plane, in agreement with inferences made on the basis of Hale's polarity laws. Even though the stratification is convectively stable at this depth, convective undershoot from above introduces strong local fluctuations in the magnetic field, without however destroying its large-scale organization.

What is truly remarkable is that this large-scale toroidal field undergoes fairly regular solar-like polarity reversals on multi-decadal timescales. This is shown on Fig. 3.22, in the form of a time–latitude diagram of the zonally-averaged toroidal magnetic component at the core–envelope interface (top), time–radius diagram of the same at 45° in the Southern hemisphere (middle), and a time–latitude diagram of the zonally-averaged surface radial magnetic component (bottom). This simulation spans 336 yr, in the course of which 11 polarity reversals have taken place, with a mean (half-) period of almost exactly 30 yr here. Examination of the top panel on Fig. 3.22 reveals a tendency for equatorward migration in the course of each cycle. The middle panel reveals that the cycles begin well within the convective envelope,

---

[17] Various animations pertaining to this simulation can be viewed on the course web-page.

3.5 Global MHD Simulations

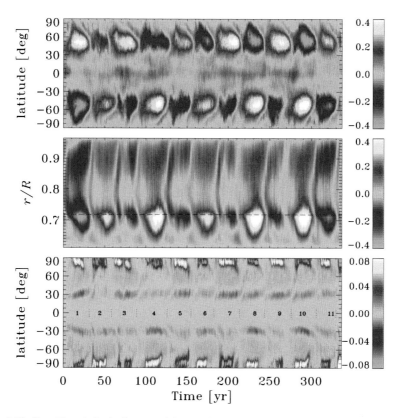

Fig. 3.22 *Top*: Time–latitude diagram of the zonally-averaged toroidal magnetic field at a depth corresponding to the interface between the convectively unstable layers and the underlying stable region. Note the antisymmetry of the large-scale field about the equatorial plane, the regular polarity reversals fairly synchronous across hemispheres, and the hint of equatorward migration of the toroidal field in the course of each cycle. The *middle panel* shows a time–radius cut of the same at mid-latitudes in the Southern hemisphere, with the *dashed line* marking the base of the convecting shell. Although the toroidal field pervades the whole convecting layers, it becomes strongly concentrated immediately beneath the convecting layers. The *bottom panel* shows a time–latitude diagram for the surface radial component. The latter reveals a well-defined axisymmetric dipole moment, oscillating essentially in phase with the deep-seated toroidal component. The color scale codes the magnetic field strength, measured in tesla. Compare to Fig. 3.4. From Racine et al. (2011) by permission of the AAS.

with later accumulation and intensification of the toroidal component at and immediately beneath the core–envelope interface, where the toroidal field strength can peak at almost half a tesla for the stronger cycles. The bottom panel on Fig. 3.22 also shows the existence of a well-defined dipole moment aligned with the rotational axis, reversing polarity approximately in phase with the deep-seated toroidal component.

Note also how, despite significant fluctuations in the amplitude and timing of the cycle in each hemisphere, in general both hemispheres remain well-synchronized throughout the whole simulation. This is all extremely solar-like!

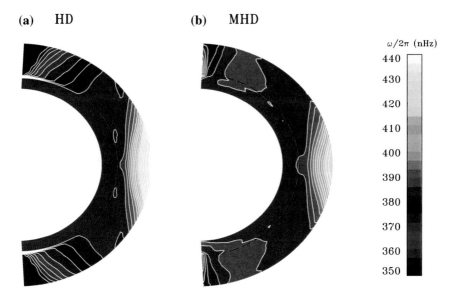

Fig. 3.23 Zonally averaged angular velocity profiles plotted over a meridional plane, in **a** a hydrodynamical (unmagnetized) version of the simulation and **b** with the equivalent profiles in the MHD simulation. The rotation axis is oriented vertically, and the dashed circular arc indicates the base of the convectively unstable layers ($r/R = 0.718$). Note the sharpness of the transition in angular velocity at this depth in the unmagnetized simulation and the much reduced pole-to-equator contrast in the MHD simulation.

What kind of dynamo could this be? To answer this question we need first to look in more detail at the flow fields. Figure 3.23 shows zonally averaged angular velocity profiles plotted in meridional $[r, \theta]$ planes. Panel a shows the corresponding profile for an unmagnetized version of the simulation of Fig. 3.22, running in the same parameter regime and subjected to the same thermal forcing. The latter is characterized by a differential rotation profile that shows a number of helioseismically-inferred solar-like features, notably equatorial acceleration and polar deceleration with a $\sim 25\%$ contrast, with near-radial Ω-isocontours at mid- to high latitudes. At these latitudes the latitudinal differential rotation vanishes abruptly in the stable layers, the transition taking place across a thin, tachocline-like shear layer coinciding with the base of the convecting layers. The most non-solar feature is the strong shear region prominent at low-latitudes within the convecting layers. The tendency for alignment of Ω-isocontours with with the rotation axis is a reflection of the Taylor–Proudman theorem, which states that in rotation-dominated systems (Coriolis term dominating over inertial and viscous terms on the RHS of Eq. (1.80)), the flow velocity cannot vary in the direction parallel to the rotation axis. In the MHD version of the simulation (Fig. 3.23b) equatorial acceleration remains, but the pole-to-equator angular velocity contrast falls to about one third of what is observed on the sun. This suggests that magnetically-mediated reduction of the large-scale flows is an important dynamo amplitude-limiting mechanism in this simulation, an inference supported by the fact

3.5 Global MHD Simulations

that significant torsional oscillations are also present, varying on the same $\sim 30\,\mathrm{yr}$ period as the large-scale magnetic field.

How about the regeneration of the large-scale poloidal component? In mean-field electrodynamics this takes place through the production of a mean electromotive force associated with the small-scale fluctuating flow and magnetic field. In the simulation considered here, the presence of a well-defined axisymmetric magnetic component suggests the definition of the "mean" flow and magnetic field through zonal averages:

$$\langle \boldsymbol{u}\rangle(r,\theta,t) = \frac{1}{2\pi}\int_0^{2\pi} \boldsymbol{u}(r,\theta,\phi,t)\,\mathrm{d}\phi, \qquad (3.68)$$

$$\langle \boldsymbol{B}\rangle(r,\theta,t) = \frac{1}{2\pi}\int_0^{2\pi} \boldsymbol{B}(r,\theta,\phi,t)\,\mathrm{d}\phi. \qquad (3.69)$$

The small-scale components then become *defined* by subtracting these mean quantities from the total flow and magnetic field vectors returned by the simulation:

$$\boldsymbol{u}'(r,\theta,\phi,t) = \boldsymbol{u}(r,\theta,\phi,t) - \langle \boldsymbol{u}\rangle(r,\theta,t), \qquad (3.70)$$
$$\boldsymbol{B}'(r,\theta,\phi,t) = \boldsymbol{B}(r,\theta,\phi,t) - \langle \boldsymbol{B}\rangle(r,\theta,t) \qquad (3.71)$$

(compare with Eq. (3.4)!). With $\boldsymbol{u}'$ and $\boldsymbol{B}'$ so defined, it is then a simple matter to calculate the mean emf directly via Eq. (3.8). With the mean emf and mean magnetic field in hand, one can then, at each grid point (r_k,θ_l) in the $[r,\theta]$ plane, *calculate the components of the α-tensor through a simple least-square fit of the nine pairs of time series $\{\mathcal{E}_i(t),\langle B\rangle_j(t)\}$*. This is carried out by minimizing a residual defined as:

$$R_{ij}(r_k,\theta_l,t) = \mathcal{E}_i(r_k,\theta_l,t) - \alpha_{ij}(r_k,\theta_l)\langle B\rangle_j(r_k,\theta_l,t),\ i,j = \{r,\theta,\phi\}. \qquad (3.72)$$

The result of this procedure is shown on Fig. 3.24a, for the $\alpha_{\phi\phi}$ component, the primary source of large-scale poloidal fields in conventional mean-field models of the solar cycle. This components reproduces many of the features "predicted" by mean-field theory and also uncovered in local simulations of MHD turbulence with an imposed large-scale field: an α-effect antisymmetric about the equatorial plane and positive in the Northern hemisphere, with a sign change in the bottom portion of the convecting layers. Moreover, the $\alpha_{\phi\phi}$ tensor component is found to be proportional to the negative of kinetic helicity (plotted on panel c) to a good first approximation, in agreement with the prediction from mean-field theory in the SOCA approximation (cf. Eq. (3.16)). In fact, reconstructing the α-tensor via Eq. (3.19), as shown on Fig. 3.24b, reveals that the current helicity (Fig. 3.24d) plays only a minor role here, with the kinetic helicity setting the spatial variations of the α-tensor. This good agreement is quite surprising because the turbulence in this simulation is strongly inhomogeneous, strongly anisotropic, and is strongly influenced by the magnetic field, which violates the underlying assumptions on which SOCA is based.

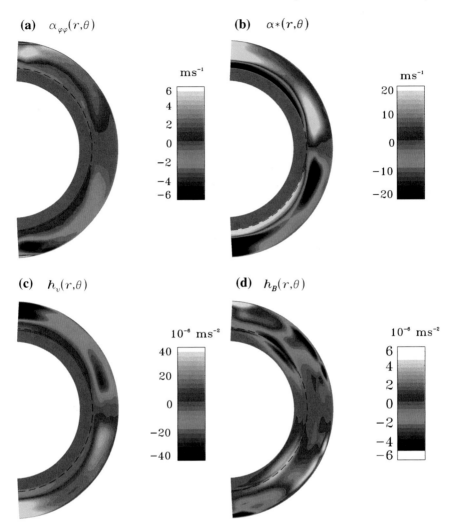

Fig. 3.24 a The $\phi\phi$ component of the α-tensor, as extracted from the numerical simulation. Note the hemispheric antisymmetry, and the sign change near the base of the convectively unstable shell. **b** The same, this time reconstructed according to Eq. (3.19), using the zonally-averaged kinetic helicity and current helicity profiles also extracted from the simulation and plotted on panels **c** and **d**, respectively. The two $\alpha_{\phi\phi}$ components are remarkably similar here, even though this simulation is operating in a regime where the SOCA approximation is not expected to hold. From Racine et al. (2011) by permission of the AAS.

The combination of a well-defined mean axisymmetric differential rotation and mean turbulent electromotive force producing a strong $\alpha_{\phi\phi}$ tensor component would suggest that this simulation may be operating as the $\alpha\Omega$ dynamos considered earlier (Sect. 3.2). Moreover, the production of a positive dipole moment from a positive toroidal field in the Northern hemisphere (see Fig. 3.21) is indeed what one would associate in mean-field theory with a positive $\alpha_{\phi\phi}$ in the Northern hemisphere. On the other hand, in this specific simulation the α-tensor has all nine of its components showing comparable amplitudes, with significant turbulent pumping contributing to the spatiotemporal evolution of the large-scale magnetic field. This also reflects the fact that the *poloidal* component of the electromotive force is quite significant, having in fact here a magnitude comparable to the shearing term arising from differential rotation. This would then suggest the $\alpha^2\Omega$ mode of dynamo action, although of a somewhat peculiar nature because here the large- and small-scale inductive contribution turn out to *oppose* each other throughout a large portion of the convection zone. This situation is not unique to this one specific simulation, having been noted already in other similar MHD simulations of solar/stellar convection using different modelling approaches.[18]

3.6 Local MHD Simulations

Throughout this chapter we have encountered various dynamo models of the solar cycle, including a dynamically correct MHD simulation, each in their own way producing a large-scale magnetic field undergoing polarity reversals in a manner not too dissimilar to what is observed on the sun. We also argued in Sect. 2.7 that the small-scale magnetic field observed at the solar surface could well be produced by local, fast dynamo action powered by the vigorous surface and subsurface turbulent convection. Are we then in a situation where *two* distinct dynamos are operating in the solar convection zone, one producing the large-scale magnetic component traditionally associated with the solar cycle, and a second powering surface magnetism away from active regions?

Observational support for the idea of a local, subsurface dynamo mechanism can be found in the fact that a tally of observed solar surface magnetic structures reveals a frequency distribution taking the form of a power law spanning over five orders of magnitude in magnetic flux, with logarithmic slope -1.85. This is a remarkable instance of scale invariance of the type most readily produced by fast dynamo action (cf. Fig. 2.19 and accompanying discussion). However, the presence of a scale-free distribution of magnetic structures at the solar surface does not *necessarily* imply fast dynamo action. Convective turbulence can reprocess magnetic flux originating elsewhere, be it deep in the convective envelope or through the decay of active regions. Likewise, surface flow can lead to the merging of magnetic structures, a process that is also self-similar and that can therefore, in principle, lead to a scale-free size distribution.

At and below the solar photosphere, the density scale height is small, convective velocities can approach the local sound speed, and radiation plays an important role in surface cooling; the anelastic approximation is no longer a viable option, and the MHD equations (1.79)–(1.82) must be solved in their fully compressible regime

---

[18] See papers by Brown et al. (2010) cited in the bibliography at the end of this chapter.

and with proper treatment of ionization and radiative transfer. Moreover, the spatial resolution must be high enough to capture granulation, the dominant surface convection pattern with a typical length scale of $\sim 10^3$ km, with intergranular downflow lanes an order of magnitude smaller (at least). This is well beyond the reach of the type of global dynamo simulations just discussed, but is accessible to *local* simulations modelling just a small portion of the convective envelope. Figure 3.25 gives an example of such a simulation, in the form of a snapshot of the "photosphere" showing emergent intensity (grayscale), on which are superimposed ± 0.1 T isocontours of vertical field strength. This is a $1000 \times 1000 \times 490$ compressible MHD simulation, with 48 km horizontal resolution and going from the upper photosphere (optical depth 0.01) down to 20 Mm in depth. In this specific simulation, a uniform horizontal magnetic field of strength 0.1 T is advected into the simulation domain through the bottom boundary.

The granulation pattern is quite obvious on Fig. 3.25, with cells of hotter (brighter) rising fluids delineated by darker, narrow downflow lanes of colder fluid, the telltale signature typical of thermally-driven convection in a stratified environment. At this (relatively) late time in the simulation, some of the magnetic flux injected at the base has reached the photosphere. This magnetic field is swept horizontally by the granular flow and accumulates in intergranular lanes. Here the combination of flux emergence and surface evolution has managed to produce a few flux concentrations sufficiently large to impede convection and form the simulation's equivalent of so-called pores (akin to small sunspots without penumbrae), where the strength of the vertical magnetic field reaches a few tenths of tesla. Notice also how many of the smaller magnetic structures, of size comparable to the width of downflow lanes and often seen where multiple downflow lanes meet, show an intensity excess above and beyond what is observed in the center of granules.

Strictly speaking, this is not a dynamo, as magnetic flux is being continuously injected through the bottom boundary. In this simulation turbulent convection is mostly "reprocessing" this magnetic flux, through the now usual mechanisms of flux expulsion, constructive and destructive folding, shearing, stretching, etc. The formation of surface flux concentrations results from the accumulation of magnetic fields in convective downflow lanes, with associated merging or cancellation depending on the relative polarities of the field elements involved. The resulting distribution of surface magnetic flux is once again a power-law, and would be very hard to distinguish from that produced exclusively by fast dynamo action driven by turbulent convection, as discussed earlier in Sect. 2.7. This highlights the difficulty to distinguish observationally local subsurface dynamo action from reprocessing of flux generated elsewhere, be it by a deep-seated large-scale dynamo or through the decay of active regions. Whether there is one or two (or more!) distinct dynamos operating in the sun remains, at this writing, an open question.

Although global MHD simulations are just beginning to yield solar-like regular cyclic global magnetic polarity reversals, they remain extremely demanding computationally, and are still a long way from producing anything resembling a toroidal flux rope, let alone a sunspot—although the formation of active regions has now been

3.6 Local MHD Simulations

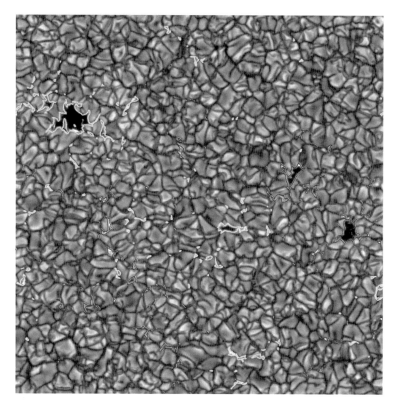

Fig. 3.25 Snapshot of the 48 × 48 Mm top "horizontal" plane of a MHD numerical simulation of thermally-driven stratified turbulent convection, with a 0.1 T uniform, horizontal magnetic field injected at the base of the simulation box (depth 20 Mm). The gray scale codes the emergent radiative flux, on which contours of constant vertical magnetic field are superimposed (*red/yellow* for $B_z = \pm 0.1$ T). Numerical simulation results kindly provided by R. F. Stein, Michigan State University.

simulated in local MHD simulations.[19] This is why the much simpler mean-field and mean-field-like cycle models described earlier in this chapter remain at this writing the favored modelling framework within which to investigate the observed characteristics of solar and stellar cycles, and in particular the origin of fluctuations in their amplitude and duration on long timescales. This is the topic to which we now turn.

---

[19] See the paper by Cheung et al. (2010) listed in the bibliography at the end of this chapter.

Bibliography

The observational literature on the solar magnetic field is immense, and many good review papers are available. One I particularly like is

Solanki, S. K., Inhester, B., & Schüssler, M.: 2006, *The solar magnetic field*, Rep. Prog. Phys., **69**, 563–668

Hale's original papers on sunspots are still well worth reading. The two key papers are:

Hale, G. E.: 1908, *On the probable existence of a magnetic field in sun-spots*, Astrophys. J., **28**, 315–343

Hale, G. E, Ellerman, F., Nicholson, S.B., & Joy, A.H.: 1919, *The magnetic polarity of sun-spots*, Astrophys. J., **49**, 153–178

The study of rising toroidal flux ropes, a proxy for the emergence of the solar internal toroidal field in the form of sunspot pairs, is a topic that has generated a voluminous literature. Among the many noteworthy contributions in this field, the following are recommended as starting points:

Moreno-Insertis, F.: 1986, *Nonlinear time-evolution of kink-unstable magnetic flux tubes in the convective zone of the sun*. Astron. & Astrophys. **166**, 291–305

Choudhuri, A. R., & Gilman, P. A.: 1987, *The influence of the Coriolis force on flux tubes rising through the solar convection zone*, Astrophys. J., **316**, 788–800

Fan, Y., Fisher, G. H., & Deluca, E. E.: 1993, *The origin of morphological asymmetries in bipolar active regions*, Astrophys. J., **405**, 390–401

D'Silva, S., & Choudhuri, A. R.: 1993, *A theoretical model for tilts of bipolar magnetic regions*, Astron. & Astrophys., **272**, 621–633

Caligari, P., Moreno-Insertis, F., & Schüssler, M.: 1995, *Emerging flux tubes in the solar convection zone. I: Asymmetry, tilt, and emergence latitude*. Astrophys. J., **441**, 886–902

The thin flux-tube approximation used in most of these calculations is due to

Spruit, H. C.: 1981, *Motion of magnetic flux tubes in the solar convection zone and chromosphere*, Astron. & Astrophys., **98**, 155–160

On the storage and stability of toroidal flux ropes below the solar convective envelope, see

Ferriz-Mas, A., & Schüssler, M.: 1994, *Waves and instabilities of a toroidal magnetic flux tube in a rotating star*, Astrophys. J., **433**, 852–866

Ferriz-Mas, A.: 1996, *On the storage of magnetic flux tubes at the base of the solar convection zone*, Astrophys. J. **458**, 802–816

Considerable effort is currently being put into doing away with the thin flux tube approximation, in order to see which of the above results remains robust, once the flux tube is no longer treated as a one-dimensional object. This is a rapidly moving field, so for the latest see the following recent on-line review:

Fan, Y.: 2009, *Magnetic fields in the solar convection zone*, Liv. Rev. Solar Phys., **6**, 4, http://solarphysics.livingreviews.org/Articles/lrsp-2009-4/

The following three recent review papers jointly offer a good overview of dynamo models of the solar cycle:

Charbonneau, P.: 2010, *Dynamo models of the solar cycle*, Liv. Rev. Solar Phys., **7**, 3, http://solarphysics.livingreviews.org/Articles/lrsp-2010-3/

Ossendrijver, M.: 2003, *The solar dynamo*, Astron. & Astrophys. Rev., **11**, 287–367

Hoyng, P.: 2003, *The field, the mean, and the meaning*, in *Advances in Non-Linear Dynamos*, Ferriz Mas, A., & Jiménez, M. M., eds., The Fluid Mechanics of Astrophysics and Geophysics, 9, Taylor & Francis, 1–36

Mean-field electrodynamics grew out of the original pioneering efforts of

Parker, E. N.: 1955, *Hydromagnetic dynamo models*. Astrophys. J., **122**, 293–314
Braginskii, S. I.: 1964, *Self-excitation of a magnetic field during motion of a highly conducting fluid*, Sov. Phys. JETP, **20**, 726–735
Steenbeck, M., & Krause, F.: 1969, *Zur Dynamotheorie stellarer und planetarer Magnetfelder. I. Berechnung sonnenähnlicher Wechselfeldgeneratoren*, Astron. Nachr., **291**, 49–84, in German

but the following three monographs are a better starting point for those wishing to dig deeper into the subject:

Moffatt, H. K.: 1978, *Magnetic Field Generation in Electrically Conducting Fluids*, Cambridge University Press
Parker, E. N.: 1979, *Cosmical Magnetic Fields: Their Origin and their Activity*, Clarendon Press, chap. 18
Krause, F., & Rädler, K.-H.: 1980, *Mean-Field Magnetohydrodynamics and Dynamo Theory*, Pergamon Press

The sketch shown on Fig.3.5 is from

Parker, E. N.: 1970, *The generation of magnetic fields in astrophysical bodies. I. The dynamo equations*, Astrophys. J., **162**, 665–673

On empirical estimates of the α-effect from numerical simulations of MHD turbulence, start with:

Pouquet, A., Frisch, U., & Léorat, J.: 1976, *Strong MHD helical turbulence and the nonlinear dynamo effect*, J. Fluid Mech., **77**, 321–354
Ossendrijver, M., Stix, M., & Brandenburg, A.: 2001, *Magnetoconvection and dynamo coefficients: Dependence of the α effect on rotation and magnetic field*, Astron. & Astrophys., **376**, 713–726
Käpylä, P. J., Korpi, M. J., Ossendrijver, M., & Stix, M.: 2006, *Magnetoconvection and dynamo coefficients. III. α-Effect and magnetic pumping in the rapid rotation regime*, Astron. & Astrophys., **455**, 401–412
Hubbard, A., Del Sordo, F., Käpylä, P. J., & Brandenburg, A.: 2009, *The α effect with imposed and dynamo-generated magnetic fields*, Mon. Not. Roy. Astron. Soc., **398**, 1891–1899

The technical literature on dynamo models of the solar cycle is truly immense. There are many hundreds of noteworthy papers out there! Those included below are just meant to be good entry points for those wishing to pursue in greater depth topics covered in this chapter. For a good overview of mean-field solar cycle models and their evolution in time, see

Lerche, I., & Parker, E. N.: 1972, *The generation of magnetic fields in astrophysical bodies. IX. A solar dynamo based on horizontal shear*, Astrophys. J., **176**, 213–223
Yoshimura, H.: 1975, *Solar-cycle dynamo wave propagation*, Astrophys. J., **201**, 740–748
Ivanova, T. S., & Ruzmaikin, A. A.: 1976, *A magnetohydrodynamic dynamo model of the solar cycle*, Sov. Astron., **20**, 227–233
Stix, M.: 1976, *Differential rotation and the solar dynamo*, Astron. & Astrophys., **47**, 243–254
Rüdiger, G., & Brandenburg, A.: 1995, *A solar dynamo in the overshoot layer: cycle period and butterfly diagram*, Astron. & Astrophys., **296**, 557–566
Moss, D., & Brooke, J.: 2000, *Towards a model for the solar dynamo*, Mon. Not. Roy. Astron. Soc., **315**, 521–533

On the impact of meridional circulation on dynamo waves, see

Bullard, E. C.: 1955, *The magnetic fields of sunspots*, Vistas in Astronomy **1**, 685–691
Choudhuri, A. R., Schüssler, M., & Dikpati, M.: 1995, *The solar dynamo with meridional circulation*, Astron. & Astrophys., **303**, L29–L32

Küker, M., Rüdiger, G., & Schultz, M.: 2001, *Circulation-dominated solar shell dynamo models with positive alpha-effect*, Astron. & Astrophys., **374**, 301–308

Roberts, P. H., & Stix, M.: 1972, *α-Effect dynamos, by the Bullard-Gellman formalism*, Astron. & Astrophys., **18**, 453–466

The meridional circulation profile described in Sect. 3.2.1 is the creation of

van Ballegooijen, A. A., & Choudhuri, A. R.: 1988, *The possible role of meridional flows in suppressing magnetic buoyancy*, Astrophys. J., **333**, 965–977

On α-quenching, standard versus catastrophic and related dynamical issues:

Blackman, E. G., & Field, G. B.: 2000, *Constraints on the magnitude of α in dynamo theory*, Astrophys. J., **534**, 984–988

Cattaneo, F., & Hughes, D. W.: 1996, *Nonlinear saturation of the turbulent α effect*, Phys. Rev. E, **54**, R4532–R4535

Durney, B. R., De Young, D. S., & Roxburgh, I. W.: 1993, *On the generation of the large-scale and turbulent magnetic fields in solar-type stars*, Solar Phys. **145**, 207–225

Rüdiger, G., & Kichatinov, L. L.: 1993, *Alpha-effect and alpha-quenching*, Astron. & Astrophys., **269**, 581–588

Cattaneo, F., & Hughes, D. W.: 2009, *Problems with kinematic mean field electrodynamics at high magnetic Reynolds numbers*, Mon. Not. Roy. Astron. Soc., **395**, L48–L51

On interface dynamos, see

Charbonneau, P., & MacGregor, K. B.: 1996, *On the generation of equipartition-strength magnetic fields by turbulent hydromagnetic dynamos*, Astrophys. J. Lett., **473**, L59–L62

MacGregor, K. B., & Charbonneau, P.: 1997, *Solar interface dynamos. I. Linear, kinematic models in Cartesian geometry*, Astrophys. J., **486**, 484–501

Parker, E. N.: 1993, *A solar dynamo surface wave at the interface between convection and nonuniform rotation*, Astrophys. J., **408**, 707–719

Petrovay, K., & Kerekes, A.: 2004, *The effect of a meridional flow on Parker's interface dynamo*, Mon. Not. Roy. Astron. Soc., **351**, L59–L62

Tobias, S. M.: 1996, *Diffusivity quenching as a mechanism for Parker's surface dynamo*, Astrophys. J., **467**, 870–880

on the energetics of thin layer dynamos:

Steiner, O., & Ferriz-Mas, A.: 2005, *Connecting solar radiance variability to the solar dynamo with the virial theorem*, Astron. Nachr., **326**, 190–193

What is now referred to as Babcock-Leighton solar-cycle models goes back to the following three seminal papers by H. W. Babcock and R. B. Leighton:

Babcock, H. W.: 1961, *The topology of the Sun's magnetic field and the 22-year cycle*, Astrophys. J., **133**, 572–587

Leighton, R. B., 1964, *Transport of magnetic fields on the Sun*, Astrophys. J., **140**, 1547–1562

Leighton, R. B.: 1969, *A magneto-kinematic model of the solar cycle*, Astrophys. J., **156**, 1–26

Although some details of the model are different, the 2D surface simulations described in Sect. 3.3.1 basically follow

Wang, Y.-M., Nash, A. G., & Sheeley, Jr., N. R.: 1989, *Magnetic flux transport on the sun*, Science, **245**, 712–718

Wang, Y.-M., & Sheeley, Jr., N. R.: 1991, *Magnetic flux transport and the sun's dipole moment - New twists to the Babcock-Leighton model*, Astrophys. J., **375**, 761–770

but on this general topic of surface magnetic flux evolution, see also:

Schrijver, C. J., Title, A. M., van Ballegooijen, A. A., Hagenaar, H. J., & Shine, R. A.: 1997, *Sustaining the quiet photospheric network: the balance of flux emergence, fragmentation, merging, and cancellation*, Astrophys. J., **487**, 424–436

Schrijver, C. J.: 2001, *Simulations of the photospheric magnetic activity and outer atmospheric radiative losses of cool stars based on characteristics of the solar magnetic field*, Astrophys. J., **547**, 475–490

Schrijver, C. J., & Title, A. M.: 2001, *On the formation of polar spots in Sun-like stars*, Astrophys. J., **551**, 1099–1106

Schrijver, C. J., De Rosa, M. L., & Title, A. M.: 2002, *What is missing from our understanding of long-term solar and heliospheric activity?*, Astrophys. J., **577**, 1006–1012

Baumann, I., Schmitt, D., Schüssler, M., & Solanki, S. K.: 2004, *Evolution of the large-scale magnetic field on the solar surface: a parameter study*, Astron. & Astrophys., **426**, 1075–1091

The formulation of the Babcock-Leighton solar cycle model of Sect. 3.3 is identical to

Charbonneau, P., St-Jean, C., & Zacharias, P.: 2005, *Fluctuations in Babcock-Leighton dynamos. I. Period doubling and transition to chaos*, Astrophys. J., **619**, 613–622

For different modelling approaches, see

Wang, Y.-M., Sheeley, Jr., N. R., & Nash, A. G.: 1991, *A new solar cycle model including meridional circulation*, Astrophys. J., **383**, 431–442

Durney, B. R.: 1995, *On a Babcock-Leighton dynamo model with a deep-seated generating layer for the toroidal magnetic field*, Solar Phys., **160**, 213–235

Dikpati, M., & Charbonneau, P.: 1999, *A Babcock-Leighton flux transport dynamo with solar-like differential rotation*, Astrophys. J., **518**, 508–520

Nandy, D., & Choudhuri, A. R.: 2001, *Toward a mean field formulation of the Babcock-Leighton type solar dynamo. I. α-coefficient versus Durney's double-ring approach*, Astrophys. J., **551**, 576–585

Guerrero, G., & de Gouveia Dal Pino, E. M.: 2008, *Turbulent magnetic pumping in a Babcock-Leighton solar dynamo model*, Astron. & Astrophys., **485**, 267–273

Muñoz-Jaramillo, A., Nandy, D., Martens, P. C. H., & Yeates, A. R.: 2010, *A double-ring algorithm for modeling solar active regions: unifying kinematic dynamo models and surface flux-transport simulations*, Astrophys. J. Lett., **720**, L20–L25

On the "tachocline α-effect" dynamo model described in Sect. 3.4.1, and associated stability analyses, begin with:

Dikpati, M., & Gilman, P. A.: 2001, *Flux-transport dynamos with α-effect from global instability of tachocline differential rotation: a solution for magnetic parity selection in the Sun*, Astrophys. J., **559**, 428–442

Dikpati, M., Gilman, P. A., & Rempel, M.: 2003, *Stability analysis of tachocline latitudinal differential rotation and coexisting toroidal band using a shallow-water model*, Astrophys. J., **596**, 680–697

Gilman, P. A., & Fox, P. A.: 1997, *Joint instability of latitudinal differential rotation and toroidal magnetic fields below the solar convection zone*, Astrophys. J., **484**, 439–454

and for the "flux tube α-effect" dynamo model of Sect. 3.4.2, and associated stability analyses, try first:

Ferriz-Mas, A., Schmitt, D., & Schüssler, M.: 1994. *A dynamo effect due to instability of magnetic flux tubes*. Astron. & Astrophys., **289**, 949–956

Ossendrijver, M. A. J. H.: 2000, *The dynamo effect of magnetic flux tubes*, Astron. & Astrophys., **359**, 1205–1210

On the numerical simulations of global 3D MHD convection in thick, rotating stratified spherical shells, begin with

Brun, A. S., Miesch, M. S., & Toomre, J.: 2004, *Global-scale turbulent convection and magnetic dynamo action in the solar envelope*, Astrophys. J., **614**, 1073–1098

Browning, M. K., Miesch, M. S., Brun, A. S., & Toomre, J.: 2006, *Dynamo action in the solar convection zone and tachocline: pumping and organization of toroidal fields*, Astrophys. J. Lett., **648**, L157–L160

Brown, B. P., Browning, M. K., Brun, A. S., Miesch, M. S., & Toomre, J.: 2010, *Persistent magnetic wreaths in a rapidly rotating Sun*, Astrophys. J., **711**, 424–438

Brown, B. P., Miesch, M. S., Browning, M. K., Brun, A. S., & Toomre, J.: 2011, *Magnetic cycles in a convective dynamo simulation of a young solar-type star*, Astrophys. J., **731**, id. 69

as well as the following two recent review articles:

Miesch, M. S.: 2005, *Large-scale dynamics of the convection zone and tachocline*, Living Reviews Solar Phys., **2**, 1, http://solarphysics.livingreviews.org/Articles/lrsp-2005-1/

Miesch, M. S., & Toomre, J.: 2009, *Turbulence, Magnetism, and Shear in Stellar Interiors*, Ann. Rev. Fluid Mech., **41**, 317–345

See also the fascinating results presented in

Cline, K. S., Brummell, N. H., & Cattaneo, F.: 2003, *Dynamo action driven by shear and magnetic buoyancy*, Astrophys. J., **599**, 1449–1468

Käpylä, P. J., Korpi, M. J., Brandenburg, A., Mitra, D., & Tavakol, R.: 2010, *Convective dynamos in spherical wedge geometry*, Astron. Nachr., **331**, 73–81

The production of solar-like magnetic cycles in such simulations is a recent breakthrough. The simulation results presented in Sect. 3.5 are taken from

Ghizaru, M., Charbonneau, P., & Smolarkiewicz, P. K.: 2010, *Magnetic cycles in global large-eddy simulations of solar convection*, Astrophys. J. Lett., **715**, L133–L137

Racine, É., Charbonneau, P., Ghizaru, M., Bouchat, A., & Smolarkiewicz, P. K.: 2011, *On the mode of dynamo action in a global large-eddy simulation of solar convection*, Astrophys. J., **735**, id. 46

These simulations were computed with the MHD version, developed at the Université de Montréal, of the general purpose hydrodynamical simulation code EULAG; on the latter,

Prusa, J. M., Smolarkiewicz, P. K., & Wyszogorodzki, A. A.: 2008, *EULAG, a computational model for multi-scale flows*, Comp. Fluids, **37**, 1193–1207

as well as the EULAG web-page:

http://www.mmm.ucar.edu/eulag/

The numerical simulation results displayed on Fig. 3.25 is publicly available at:

http://steinr.pa.msu.edu/~bob/data.html

Explanatory notes describing the simulation framework are also provided there, and discussed in greated detail in

Stein, R. F., Lagerfjärd, A., Nordlund, Å., & Georgobiani, D.: 2011, *Solar flux emergence simulations*, Solar Phys., **268**, 271–282

In a similar vein, do not miss:

Cheung, M. C. M., Rempel, M., Title, A. M., & Schüssler, M.: 2010, *Simulation of the formation of a solar active region*, Astrophys. J. **720**, 233–244

On the observational measurements and characterization of small-scale solar surface magnetic structures, and the potential implications for dynamo processes, see

Parnell, C. E., DeForest, C. E., Hagenaar, H. J., Johnston, B. A., Lamb, D. A., & Welsch, B. T.: 2009, *A power-law distribution of solar magnetic fields over more than five decades in flux*, Astrophys. J. **698**, 75–82

and references therein. A simple diffusion-limited aggregation model, producing power-law distributions of magnetic structures with logarithmic slope comparable to observational inferences, is presented in

Crouch, A. D., Charbonneau, P., & Thibault, K.: 2007, *Supergranulation as an emergent length scale*, Astrophys. J., **662**, 715–729

Chapter 4
Fluctuations, Intermittency and Predictivity

> *It is nice to know that the computer understands the problem,
> but I would like to understand it too.*
> Attributed to E.P. Wigner

Given that the basic physical mechanism(s) underlying the operation of the solar cycle are not yet agreed upon, attempting to understand the origin of the observed *fluctuations* of the solar cycle may appear to be a futile undertaking. Nonetheless, work along these lines continues at full steam in part because of the high stakes involved: the solar radiative output and frequencies of all eruptive phenomena relevant to space weather are strongly modulated by the amplitude of the solar cycle; varying levels of solar activity may contribute significantly to climate change; and certain aspects of the observed fluctuations may actually hold important clues as to the physical nature of the dynamo process.

We first briefly review some classical solar cycle fluctuation patterns, as inferred from the sunspot number time series (Sect. 4.1). With an eye on reproducing these patterns, we then study the response of some of the basic dynamo models considered in the preceding chapter to stochastic forcing (Sect. 4.2), dynamical nonlinearities (Sect. 4.3), and time delays (Sect. 4.4). We then examine how the interaction of some of these modulation mechanisms can lead to intermittency (Sect. 4.5), and close with a brief survey of the current status of model-based solar cycle prediction schemes (Sect. 4.6).

4.1 Observed Patterns of Solar Cycle Variations

4.1.1 Pre-Telescopic and Early Telescopic Sunspot Observations

Until the beginning of the twentieth century, the story of the solar activity cycle is coincident with the story of *sunspots*. Appearing as dark blemishes on the bright solar disk, only the largest sunspots can be visible to the naked-eye under suitable viewing conditions, for example when the sun is partially obscured by clouds or mist, particularly at sunrise or sunset. Numerous such sightings exist in the historical records, starting with Theophrastus (374–287 B.C.) in the fourth century B.C. However, the most extensive pre-telescopic records are found in the far east, especially in the official records of the Chinese imperial courts, starting in 165 B.C. In December 1128 the monks authoring the Worcester Chronicles also left us with the first known sunspot drawing of what must have been two exceptionally large sunspots, since their umbrae and penumbrae could be visually distinguished. A fascinating pre-telescopic sunspot sighting is certainly that of 28 May 1607 by none other than Johannes Kepler (1571–1630). Kepler had been observing the sun for over a month using his *camera obscura* projection technique, basically a pinhole camera. He was hoping to detect a transit of Mercury across the solar disk, as predicted by extant planetary ephemerides. On May 28, through a short-lived break in cloud cover, he did notice a small black spot on the solar disk, and concluded that he was indeed seeing Mercury in transit. It did not take long before he came to realize his mistake.

At the end of the first decade of the seventeenth century, the "discovery" of sunspots was definitely in the air. Within less than a year of one another, four observers turned the newly invented astronomical telescope toward the sun, and independently noted the existence of sunspots. They were Johann Goldsmid (1587–1616, a.k.a. Fabricius) in Holland, Thomas Harriot (1560–1621) in England, Galileo Galilei (1564–1642) in Italy, and the Jesuit Christoph Scheiner (1575–1650) in Germany. Fabricius was the first to publish his results in 1611, and to correctly interpret the apparent motion of sunspots in terms of axial rotation of the sun. Like Harriot, Fabricius and his father (the then-well-known astronomer David Fabricius) first observed sunspots directly through their telescope shortly after sunrise or before sunset. The harrowing account of their observations is worth quoting: (excerpt from the translation in the paper by W.M. Mitchell cited at the end of this chapter):

> "... Having adjusted the telescope, we allowed the sun's rays to enter it, at first from the edge only, gradually approaching the center, until our eyes were accustomed to the force of the rays and we could observe the whole body of the sun. We then saw more distinctly and surely the things I have described [sunspots]. Meanwhile clouds interfered, and also the sun hastening to the meridian destroyed our hopes of longer observations; for indeed it was to be feared that an indiscreet examination of a lower sun would cause great injury to the eyes, for even the weaker rays of the setting or rising sun often inflame the eye with a strange redness, which may last for two days, not without affecting the appearance of objects."

Galileo and Scheiner, however, were the most active in using sunspots to attempt to infer physical properties of the sun. To Galileo belongs the credit of making a convincing case that sunspots are indeed features of the solar surface, as opposed to intra-Mercurial planets, as advocated initially by Scheiner. But the Jesuit astronomer did detect and measure, after years of careful observations of sunspots' apparent motion on the solar disk, the slight inclination of the sun's equatorial plane with respect to the ecliptic. But as for the physical nature of sunspots, even the universally opinionated Galileo remained unusually cautious, tentatively granting them the status of clouds in the solar atmosphere.

In the three subsequent centuries sunspots became dark mountains peaking through luminous cloud layers, then orifices allowing a view of the dark solar surface, then giant cyclones until, as we already discussed in Sect. 3.1, their magnetic nature was firmly established by Hale and collaborators. But this did not prevent sunspot observers to make some striking discoveries in the three centuries separating Hale from Galileo.

4.1.2 The Sunspot Cycle

Early sunspot observers noted the curious fact that sunspots rarely appear outside of a latitudinal band of about $\pm 40°$ centered about the solar equator, but otherwise failed to discover any clear pattern in the appearance and disappearance of sunspots. This fell to the German amateur astronomer Samuel Heinrich Schwabe (1789–1875), who in 1843, after a 17-years telescopic hunt for intra-mercurial planets, announced the existence of a decadal periodicity in the average number of sunspots visible on the sun.

Much impressed by Schwabe's discovery, the Swiss astronomer Rudolf Wolf (1816–1893) launched in a life-long quest for sunspot data and drawings from previous centuries, with the aim of tracking the sunspot cycle all the way back to the beginning of the telescopic era. Faced with the daunting task of comparing sunspot observations carried out by many different astronomers using various instruments and observing techniques, Wolf defined a *relative sunspot number* (r) as follows:

$$r = k(f + 10g), \tag{4.1}$$

where g is the number of sunspot groups visible on the solar disk, f is the number of individual sunspots (including those distinguishable within groups), and k is a correction factor that varies from one observer to another (with $k = 1$ for Wolf's own observations). This definition is still in used today, but r is now officially called the international sunspot number, although the names Wolf (or Zürich) sunspot number are still in common usage. Wolf succeeded in reliably reconstructing the variations in sunspot number as far as the 1755–1766 cycle, which has since been known conventionally as "Cycle 1", with all subsequent cycles numbered consecutively thereafter; at this writing (December 2010), we are just coming out of the unusually extended minimal activity phase delineating cycle 23 from the upcoming cycle 24.

Figure 4.1 shows two time series of the relative sunspot number. The first (thin black line) is the monthly-averaged value of r as a function of time, and the thick red line is a 13-month running mean of the same. The amplitude, duration and even shape of sunspots cycles can vary substantially from one cycle to the next. The period, in particular, ranges from 9 (cycle 2) to 14 years (cycle 4). Moreover, we have already seen (Sect. 3.1) that from a physical—rather than botanical—standpoint, the full period of the underlying magnetic cycle is twice that of the sunspot cycle. Yet, because the near totality of phenomena defining solar activity are unaffected by the magnetic polarity of the sun's large-scale magnetic field, and also perhaps because

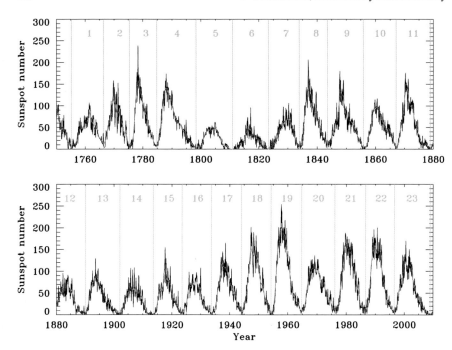

Fig. 4.1 Two time series of the celebrated Wolf Sunspot Number. The *thin black line* is the monthly-averaged sunspot number, and the *thick red line* a 13-month running mean thereof. On the basis of the latter, one can calculate a mean cycle peak amplitude of sunspot number SSN = 115 with standard deviation ±40, and a mean period 10.8 yr with standard deviation ±1.6 yr. Cycles are defined from one minimum to the next, and are numbered following Wolf's convention. These and other related data are publicly available at the Solar Influences Data Analysis Center in Brussels, Belgium, http://sidc.oma.be.

astronomers are creatures of tradition, even a century after Hale's discovery of the sunspot polarity law it remains customary to speak of the "11 year solar cycle".

4.1.3 The Butterfly Diagram

To the striking cyclic pattern uncovered by Schwabe was soon added an equally striking *spatial* regularity. In 1858, Gustav Spörer (1822–1895) and Richard Carrington (1826–1875) independently pointed out that sunspots are observed at relatively high ($\sim 40°$) heliocentric latitudes at the beginning of a sunspot cycle, but are seen at lower and lower latitudes as the cycle proceeds, until at the end of the cycle they are observed mostly near the equator, at which time spots announcing the onset of the next cycle begin to appear again at $\sim 40°$ latitude.

This is illustrated on Fig. 4.2, in the form of a *butterfly diagram* for the time period 1875–2010. The construction of a sunspot butterfly diagram, first carried out

4.1 Observed Patterns of Solar Cycle Variations 157

Fig. 4.2 A sunspot butterfly diagram, showing the equatorward migration of sunspot latitudes in the course of each cycle. The sunspot number peaks about midway through the equatorward migration. Public domain data and graphics courtesy of David Hathaway, ASA/MSFC, http://solarscience.msfc.nasa.gov/images/bfly.gif.

in 1904 by the husband-and-wife team of Annie Maunder (1868–1947) and E. Walter Maunder (1851–1928), proceeds as follows: one begins by laying a coordinate grid on (for example) a solar white light image, with, as in the case of geographic coordinates on Earth, the rotation axis defining the North–South direction. The visible solar disk is then divided in latitudinal strips of constant projected area, and for each such strip the percentage of the area covered by sunspots and/or active regions is calculated and color coded. This defines a one-dimensional (latitudinal) array describing the average sunspot coverage at one time. By repeating this procedure at constant time intervals and stacking the arrays one besides the other, one obtains a two-dimensional image of average sunspot coverage as a function of heliospheric latitude (vertical axis) and time (horizontal axis).

The absence of sunspots at high latitudes ($\gtrsim 40°$) at any time during the cycle, and the equatorward drift of the sunspot distribution as the cycle proceeds from maximum to minimum, are both particularly striking on such a diagram. Note how the latitudinal distribution of sunspots is never exactly the same, and how for certain cycles there exists a significant North–South asymmetry in the hemispheric distributions. Note also how, for most cycles, spots from each new cycle begin to appear at mid-latitudes while spots from the preceding cycle can still be seen near the equator,[1] and how sunspots are almost never observed within a few degrees in latitude of the equator. Sunspot maximum (1991, 1980, 1969, …) occurs about midway along each butterfly, when sunspot coverage is maximal at about 15 degrees latitude. Note also how fairly good overall hemispheric synchrony is maintained, despite significant amplitude variations and occasional large lag in the beginning of the cycle in one hemisphere with respect to other, for example at the onset of the 1965–1976 cycle.

---

[1] The most recent, unusually extended activity minimum between cycles 23 and 24 was definitely unusual in this respect.

4.1.4 The Waldmeier and Gnevyshev–Ohl Rules

The sunspot number is our longest direct record of solar activity, and thus remains a favored dataset for the analysis and modelling of solar cycle fluctuations. Starting with Wolf himself, the sunspot number time series (monthly, monthly smoothed, yearly, etc.) have been analyzed in every possible manner known to statistics, nonlinear dynamics, and numerology.[2] Many otherwise serious and respectable people engaged in this type of work seem to forget that the definition of the sunspot number is largely arbitrary, and its *quantitative* relationship to the real dynamical quantity, the sun's internal magnetic field, uncertain at best.

Nonetheless, some of the various patterns so uncovered do appear to be robust, in that they do not depend too much on the manner the analysis is being carried out, and are also found in other indicators of solar activity; for example the so-called *Gleissberg cycle* refers to a $\sim 80\,\mathrm{yr}$ modulation of the overall envelope of cycle amplitudes. Some sunspot number patterns have even proven resilient enough to be upgraded to the status of empirical "rules", two of the more convincing ones being the so-called Waldmeier rule, and Gnevyshev–Ohl rule.

The *Waldmeier rule* refers to an anticorrelation observed between cycle amplitude and rise time (or duration). Starting for example from the time series of smoothed monthly sunspot number (red line on Fig. 4.1), it is straightforward to assign to each cycle n a peak amplitude A_n and a duration T_n, the latter being simply the time interval between the two minima bracketing a given cycle. Similarly, the rise time is the time interval between a minimum and the subsequent maximum. Figure 4.3a shows a correlation plot of cycle rise time and amplitude, which is characterized by a linear correlation coefficient of $r = -0.7$, definitely large enough to merit attention. A similar, though weaker anticorrelation exists between cycle amplitude and duration, a consequence of a similarly weak correlation ($r = +0.4$) existing between cycle rise time and duration (see Fig. 4.3b). The latter correlation is taken to be indicative of some level of self-similarity in the temporal unfolding of sunspot cycles. The amplitude-duration anticorrelations are intriguing, because one might have (naively) expected that high amplitude cycles should take longer to build up and also last longer, but in fact the opposite seems to hold.

The *Gnevyshev–Ohl rule* is another intriguing pattern, and is illustrated on Fig. 4.3c. Cycle peak amplitudes A_n are plotted as solid dots, versus cycle number n, following Wolf's numbering convention. Compute now a 1-2-1 running mean of cycle amplitude, i.e.,

$$\langle A_n \rangle = \frac{1}{4}(A_{n-1} + 2A_n + A_{n+1}), \qquad n = 2, 3, \ldots . \tag{4.2}$$

---

[2] Two colleagues, David J. Thomson and Werner Mende, both world-renowned experts in time series analysis, have independently remarked to me that the sunspot number time series are quite possibly the "natural" time series having produced the largest number of research journal pages per byte of actual data!

4.1 Observed Patterns of Solar Cycle Variations

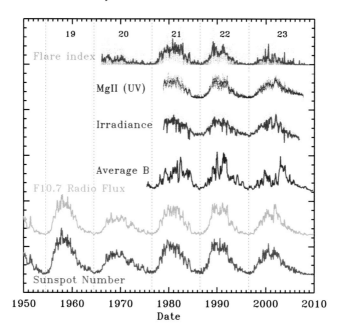

Fig. 4.4 The solar activity cycle, as measured via various proxies. From *bottom* to *top* the sunspot number, the F10.7 radio flux, the disk-averaged line-of-sight magnetic field, the total solar irradiance, the MgII index (a good proxy of ultraviolet emission), and the solar flare index. Plot constructed from public-domain data archived at NOAA (USA), available at http://www.ngdc.noaa.gov.

Maunder Minimum (cf. Fig. 4.5). On the other hand, cosmogenic radioisotopes such as ^{10}Be, whose production frequency is known to be modulated by the frequency of solar eruptive phenomena and general strength of the interplanetary magnetic field, continue to show a cyclic pattern throughout the Maunder minimum (Fig. 4.5, top panel), indicating that the cycle had actually not come to a complete standstill. The Maunder Minimum remains a real puzzle in many ways.

The cosmogenic isotope record also indicates that similar episodes of markedly reduced solar activity occurred in 1282–1342 (Wolf minimum) and 1416–1534 (Spörer minimum), and that solar activity was significantly above its modern average in the time period 1100–1250 (dubbed Medieval Maximum by Min/Max aficionados). Some recent such reconstructions (see bibliography) have in fact identified 27 grand minima in the past 11,000 years. No convincing periodicity or other temporal pattern has yet been identified in the occurrence of these grand minima. The 1798–1823 *Dalton Minimum*, spanning the low and oddly-shaped cycles 5 and 6, is sometimes categorized as a "failed Grand Minimum", although supporting arguments tend to be of a botanical flavor.

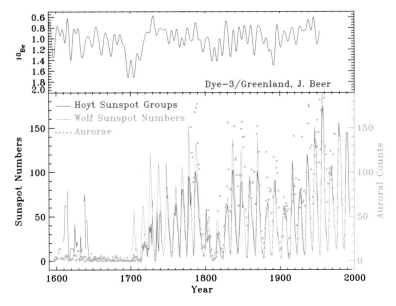

Fig. 4.5 The Maunder minimum, as seen through cosmogenic radioisotopes (*top panel*) and sunspot and auroral counts (*bottom panel*). The *thick red line* is the so-called Group Sunspot Number, a reconstruction similar to Wolf's (*thin orange line*) but deemed more reliable in the 18th century because it relies exclusively on the more easily observable sunspot groups. Beryllium 10 data courtesy of J. Beer, EAWAG/Zürich.

4.1.7 From Large-Scale Magnetic Fields to Sunspot Number

Pondering over the Maunder Minimum puzzle leads naturally into one major difficulty that plagues *any* and *all* dynamo models and MHD simulations discussed in the preceding chapter, when trying to reproduce this or that solar cycle fluctuation patterns seen in the sunspot number: what is the *quantitative* relationship between the internal large-scale magnetic field and the number of sunspots emerging at the solar surface? The process through which the dynamo-generated magnetic field produces toroidal flux ropes in the tachocline is not understood, but the few extant calculations attempting to simulate this formation process indicate that it is much more than a mere matter of toroidal field strength. The destabilization and rise of these toroidal flux ropes is also not just a matter of field strength, as the stability diagram of Fig. 3.19 already shows quite well. Once the flux rope emerges, it is not at all clear that the number of sunspots is uniquely and proportionally related to the magnetic field strength or flux in the rope; a bipolar magnetic region made up of two monolithic sunspots would contribute $2 + 10$ to the sunspot number, as defined by Eq. (4.1); with the trailing component of the bipolar region broken up into 10 small spots (say), as often observed, one gets instead a $11 + 10$ contribution to the sunspot number; this is a difference by nearly a factor of two, for the the same magnetic flux!

4.1 Observed Patterns of Solar Cycle Variations

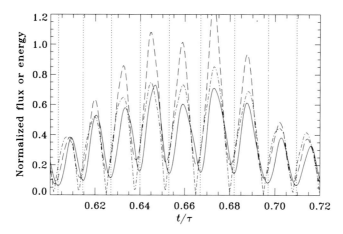

Fig. 4.6 Time series of the three different sunspot number proxies defined through Eqs. (4.3)–(4.5), constructed for the advection-dominated mean-field $\alpha\Omega$ solution of Fig. 3.10, subjected to stochastic fluctuations. The *solid line* is the total magnetic energy in the domain, the *dashed line* the magnetic energy within the tachocline, and the *dash-dotted line* the net toroidal magnetic flux within the tachocline. The three time series have been normalized so that they yield the same peak amplitude for the first cycle displayed here.

The fact remains that the sunspot number does correlate well with other more "physical" measures of the solar magnetic field, such as photospheric magnetic flux, active region magnetic flux, and the F10.7 solar radio flux (see Fig. 4.4). Until strong evidence to the contrary is presented, it is probably reasonable to assume that a more strongly magnetized sun will produce more sunspots, but it would be really surprising if that relationship were nicely and conveniently linear over a wide range of activity levels.

We are used to thinking of sunspot numbers as a proxy for the solar internal magnetic field; but starting from a dynamo solution for the solar large-scale magnetic field, we must now construct a proxy for the sunspot number! Consider the following equally "reasonable" simple proxies: the total magnetic energy, the magnetic energy within the tachocline, and the net toroidal magnetic flux in the tachocline:

$$\text{SSN}_1(t) = \int_V \boldsymbol{B}^2(r, \theta, t) \, dV \,, \tag{4.3}$$

$$\text{SSN}_2(t) = 2\pi \int_0^\pi \int_{r_c-w}^{r_c+w} \boldsymbol{B}^2(r, \theta, t) \, r^2 \sin\theta \, dr d\theta \,, \tag{4.4}$$

$$\text{SSN}_3(t) = \left| 2\pi \int_0^\pi \int_{r_c-w}^{r_c+w} B(r, \theta, t) \, r^2 \sin\theta \, dr d\theta \right| \,. \tag{4.5}$$

Even though these three proxies are closely related, they lead to sunspot-number (SSN) proxy timeseries that show some significant differences. This is illustrated on Fig. 4.6, for the advection-dominated mean-field $\alpha\Omega$ model of Sect. 3.2.11, subjected to stochastic forcing (more on this shortly, in Sect. 4.2). The plot covers a period spanning a gradual rise and decline in cycle amplitude, and the time series are normalized so as to yield the same peak amplitude for the first cycle plotted. It is reassuring to see all three proxies correlating rather well, and showing similar long-term trends in cycle amplitude and duration, yet significant differences are also apparent. Close examination of Fig. 4.6 reveals that the timing of cycle maxima and minima differs slightly from one proxy to the other, moreover in a manner dependent on the cycle amplitude. Note also how the relative differences between the three sunspot proxies depend significantly on the cycle amplitude, approaching 40% for some cycle.

For the purposes of the foregoing discussion, these differences are inconsequential, except perhaps when considering cycle prediction schemes (Sect. 4.6), where the aim is to predict sunspot number as accurately as possible. With this important caveat under the belt, we proceed with our study of solar cycle fluctuations, using the total magnetic energy as a SSN proxy.

4.2 Cycle Modulation Through Stochastic Forcing

An obvious means of producing amplitude fluctuations in dynamo models is to introduce stochastic forcing in the governing equations. Sources of stochastic "noise" certainly abound in the solar interior; large-scale flows in the convective envelope, such as differential rotation and meridional circulation, are observed to fluctuate, an unavoidable consequence of dynamical forcing by the surrounding, vigorous turbulent flow. This convection is known to produce its own small-scale magnetic field (viz. Fig. 2.23), and amounts to a form of rapidly varying zero-mean "noise" superimposed on the slowly-evolving mean magnetic field. This can be readily incorporated into dynamo models by introducing, on the RHS of the governing equations, an additional zero-mean source term localized at the surface, and varying randomly from node-to-node in latitude and from one time step to the next:

$$A(R, \theta, t) \rightarrow A(R, \theta, t) + \varrho \, \delta A, \qquad \varrho \in [-1, 1], \tag{4.6}$$

with δA a fixed amplitude, and the random number ϱ is uniformly distributed in the interval. Note that the current-free boundary condition at $r/R = 1$ for the toroidal component requires $B(R, \theta, t) = 0$, therefore we only add a perturbation to the poloidal component. This is a classical instance of *additive noise*.

In addition, the various source terms appearing in the dynamo equation will present significant temporal deviations about their azimuthal means, as a consequence of the spatiotemporally discrete nature of the physical events (e.g., cyclonic updrafts, sunspot emergences, flux rope destabilizations, etc.) whose collective effects add up to produce a mean electromotive force. The impact of these statistical fluctuations about the mean can be modeled in a number of ways. Perhaps the most straightforward

4.2 Cycle Modulation Through Stochastic Forcing

is to let the dynamo number fluctuate randomly in time about some pre-set mean value $\bar{C}_\alpha$:

$$C_\alpha \to \bar{C}_\alpha + \varrho\, \delta C_\alpha\,, \qquad \varrho \in [-1, 1]\,, \qquad \text{if}(t \bmod \tau_c) = 0\,. \tag{4.7}$$

By most statistical estimates, the expected magnitude of these fluctuations is quite large, i.e., many times the mean value, a conclusion also supported by "measurements" of the α-tensor components in numerical simulations. One typically also introduces a *coherence time* (τ_c) during which the dynamo number retains a fixed value. At the end of this time interval, this value is randomly readjusted. Depending on the dynamo model at hand, the coherence time can be related to the lifetime of convective eddies (α-effect-based mean-field models), to the decay time of sunspots (Babcock–Leighton models), or to the growth rate of instabilities (hydrodynamical shear or buoyant MHD instability-based models). Equation (4.7) represents an instance of *multiplicative noise*, since the fluctuating quantity is multiplying a source term in the governing equations, which is itself a function of the system's dependent variables.

The effect of stochastic forcing varies according to the type of dynamo model being forced, but some common trends and tendencies nonetheless emerge. In most models, stochastic forcing or noise increases both the average amplitudes and durations of cycles. It also introduces long-timescale modulations in the overall cycle amplitudes, "long" in the sense of being much longer than the assumed coherence time for the noise and/or forcing, and often significantly longer than the cycle period itself. In kinematic models, this can often be traced to the production and storage of strong magnetic fields in the low-diffusivity regions of the domain, below the core–envelope interface, where the resistive decay time of these structures can be quite long.

Figures 4.7 and 4.8 show some representative results for the advection-dominated mean-field $\alpha\Omega$ model of Sect. 3.2.11, and for the Babcock–Leighton model of Sect. 3.3.5, respectively. In both cases the total magnetic energy (red line on panels a) is used as a proxy for the sunspot number. These two specific stochastically forced solutions were selected because they exhibit a number of solar-like features, including relative ranges of variations in cycle amplitudes ($\pm \sim 40\%$ of the mean) and duration ($\pm \sim 15\%$ of the mean), amplitude modulation patterns spanning many cycles, and shorter-lived Dalton-minimum-like intervals of markedly reduced amplitude.

Both of these solutions (and many of their "cousins" computed with varying amplitude of stochastic forcing) do fairly well at reproducing Gnevyshev–Ohl-like alternating patterns of variations in cycle amplitude about their running mean. This is illustrated on panels (b) of both figures. A 1-2-1 running mean of cycle amplitudes is first computed according to Eq. (4.2), yielding the thin purple line on panels (a). This is then subtracted from the temporal sequence of cycle amplitudes, to give the "detrended" amplitudes plotted on panels (b). The gray horizontal bars flag the temporal intervals during which a regular alternance of above-and-below the mean is sustained. That such sequences should exist is in itself not surprising, in view

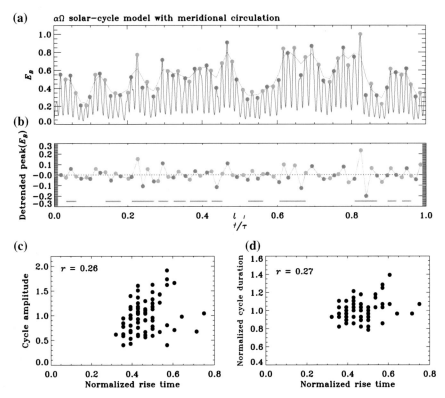

Fig. 4.7 Impact of stochastic fluctuations of the C_α dynamo number on the behavior of an advection-dominated mean-field solar cycle model including a meridional circulation. This is the solution of Fig. 3.10, with 100% forcing of the poloidal dynamo number ($\delta C_\alpha/\bar{C}_\alpha = 1$ in Eq. (4.7)). Panel **a** shows part of a time series for the magnetic energy (*red*), together with a 1-2-1 running mean of the peak amplitudes (*purple*), as defined in Eq. (4.2). Subtracting this from the temporal sequence of peak amplitudes yields the "detrended" sequence shown on panel (**b**), where odd- (even-)numbered cycles are plotted in *red* (*green*), and the *horizontal gray bars* indicate epochs where a Gnevyshev–Ohl-like pattern holds. Panels **c** and **d** are Waldmeier-rule-like correlation plots between cycle peak, rise time and duration (cf. Fig. 4.3), with cycle peak and duration normalized to their mean values over the full simulation run.

of the detrending procedure adopted here; purely random numbers would distribute themselves symmetrically about their mean, so that the Gnevyshev–Ohl patterns can materialize only by chance. What is striking here is the distribution of durations for these epochs, which can greatly exceed (especially here in the Babcock–Leighton solution) what one could rightfully expect from Poissonian statistics.

Most stochastically forced models, including the two shown on Figs. 4.7 and 4.8, do produce a positive correlation between cycle rise time and duration (cf. panels (d) and Fig. 4.3b). In the case of the mean-field model of Fig. 4.7 that correlation is too weak compared to solar, while for the Babcock–Leighton model it is too strong, but adjustment of the amplitude of stochastic forcing can readily yield a more solar-like correlation.

4.2 Cycle Modulation Through Stochastic Forcing

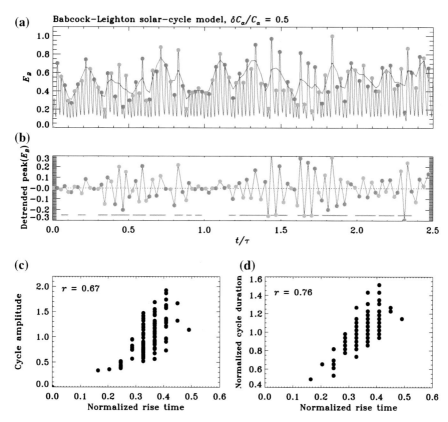

Fig. 4.8 Identical in format to Fig. 4.7, but now the parent model is the Babcock–Leighton solution of Fig. 3.17, with poloidal source term fluctuating at the level $\delta C_\alpha / \bar{C}_\alpha = 0.5$. The moderately strong positive correlation between cycle amplitudes and rise time is markedly non-solar, but the similar correlation between cycle duration and rise time is in better agreement with solar cycle data.

The situation is nowhere as good with regards to the observed anticorrelation between cycle amplitude and rise time (or duration) embodied in the Waldmeier rule (viz. Fig. 4.3a). Whether forced stochastically through the dynamo number or via additive noise in the surface layers, most of the kinematic models considered here end up producing a positive correlation (rather than an anticorrelation) between these two cycle parameters. A Waldmeier-like anticorrelation has been observed in stochastically-forced linear $\alpha\Omega$ model near criticality,[3] but this interesting result in general does not carry over to nonlinearly-saturated $\alpha\Omega$ dynamo solutions. It has also been observed in a Babcock–Leighton models subjected to stochastic perturbations imposed on the form of the meridional flow profile,[4] but again it is not clear how generic or robust this actually is. It may well be that the key to the Waldmeier rule lies at least in part with non-kinematic effects, such as the nonlinear back reaction

---
[3] See the paper by Ossendrijver & Hoyng (1996) cited in the bibliography.
[4] See the paper by Charbonneau & Dikpati (2000) cited in the bibliography.

of the dynamo-generated magnetic field on differential rotation and/or meridional circulation.

4.3 Cycle Modulation Through the Lorentz Force

The dynamo-generated magnetic field will, in general, produce a Lorentz force that will tend to oppose the driving fluid motions. This is a basic physical effect that should be included in any dynamo model. It is not all trivial to do so, however, since in a turbulent environment both the fluctuating and the mean components of the magnetic field can affect both the large-scale flow components, as well as the small-scale turbulent flow providing the Reynolds stresses powering the large-scale flows. One must thus distinguish between a number of (related) amplitude-limiting mechanisms:

1. Lorentz forces associated with the mean magnetic field *directly* affecting large-scale flows (sometimes called the "Malkus–Proctor effect").
2. Large-scale magnetic fields *indirectly* affecting large-scale flows via effects on small-scale turbulence and associated Reynolds and Maxwell stresses (sometimes called "Λ-quenching").
3. The magnetic field, large or small-scale, *directly* affecting the small-scale turbulent flow sustaining the α-effect and/or turbulent diffusivity.

The third of these we touched on already with the idea of the algebraic α-quenching (Sect. 3.2.9), but other approaches have been put forth, including so-called dynamical α-quenching, where an additional dynamical equation is introduced, describing the temporal evolution of the α-tensor component (or some related quantity such as kinetic helicity) in response to the time-varying magnetic field. Introducing magnetic backreaction on differential rotation and/or meridional circulation is a far trickier business than one might imagine, because one must then also, in principle, provide a model for the Reynolds stresses powering the large-scale flows in the solar convective envelope, as well as a procedure for computing magnetic backreaction on these. This rapidly leads into the unyielding realm of MHD turbulence, although algebraic "Λ-quenching" formulae akin to α-quenching have been proposed based on specific turbulence models. Alternately, one can add an ad hoc source term to the RHS of Eq. (1.80), designed in such a way that in the absence of the magnetic field, the desired solar-like large-scale flow is obtained. As a variation on this theme, one can simply divide the large-scale flow into two components, the first (U) corresponding to some prescribed, steady profile, and the second (U') to a time-dependent flow field driven by the Lorentz force:

$$u(x, t, B) = U(x) + U'(x, B(t)), \qquad (4.8)$$

with the (non-dimensional) governing equation for U' including only the Lorentz force and a viscous dissipation term on its RHS. This is *not* a linearisation, in that

4.3 Cycle Modulation Through the Lorentz Force

we are *not* assuming that $U' \ll U$. The time-varying flow contribution must then obey a (nondimensional) differential equation of the form

$$\frac{\partial U'}{\partial t} = \frac{\Lambda}{\mu_0 \varrho} (\nabla \times B) \times B + \mathrm{P_m} \nabla^2 U' , \qquad (4.9)$$

where time has been scaled according to the magnetic diffusion time $\tau = R^2/\eta_e$, as before. Two dimensionless parameters appear in Eq. (4.9). The first (Λ) is a numerical parameter measuring the influence of the Lorentz force. The second, $\mathrm{P_m} = \nu/\eta$, is the magnetic Prandtl number introduced earlier, and measures the relative importance of viscous versus Ohmic dissipation. When $\mathrm{P_m} \ll 1$, large velocity amplitudes in U' can be produced by the dynamo-generated mean magnetic field. This effectively introduces an additional, long timescale in the model, associated with the evolution of the magnetically-driven flow; the smaller $\mathrm{P_m}$, the longer that timescale.

The majority of studies published thus far and using this approach have only considered the nonlinear magnetic backreaction on differential rotation. This has been shown to lead to a variety of behaviors, including amplitude and parity modulation, periodic or aperiodic, as well as intermittency (more on the latter in Sect. 4.5). It has been argued that amplitude modulation in such models can be divided into two main classes:

1. Nonlinear interaction between modes of different parity, with the Lorenz force-mediated flow variations controlling the transition from one mode to another.
2. Exchange of energy between a single dynamo mode (of some fixed parity) with the flow field. This leads to quasiperiodic modulation of the basic cycle, with the modulation period controlled by the magnetic Prandtl number.

Both types of modulation can co-exist in a given dynamo model, leading to a rich overall dynamical behavior. Figure 4.9 shows two time-latitude diagrams produced by a nonlinear mean-field interface model.[5] The model is defined on a cartesian slab with a reference differential rotation varying only with depth, and includes backreaction on the differential rotation according to the procedure described above. The model exhibits strong, quasi-periodic modulation of the basic cycle, leading to epochs of strongly reduced amplitude. Note how the dynamo can emerge from such epochs with strong hemispheric asymmetries (top panel), or with a different parity (bottom panel).

The differential rotation can also be suppressed indirectly by magnetic backreaction on the *small-scale* turbulent flow that produces the Reynolds stresses driving the large-scale mean flow. Inclusion of this so-called "Λ-quenching" in mean-field dynamo models, alone or in conjunction with other amplitude-limiting nonlinearities, has also been shown to lead to a variety of periodic and aperiodic amplitude modulations, provided the magnetic Prandtl number is small.[6] This type of models stand or fall with the turbulence model used to compute the various mean-field

---

[5] For details on this model see the paper by Tobias (1997) cited in the bibliography.
[6] See, e.g., the paper by Küker et al. (1999) cited in the bibliography.

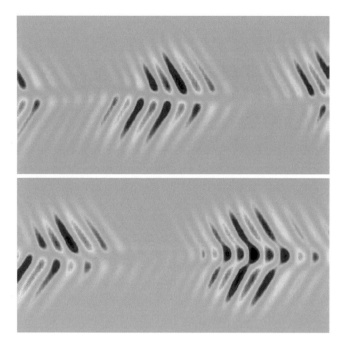

Fig. 4.9 Amplitude and parity modulation in a dynamo model including magnetic backreaction on the differential rotation. These are the usual time-latitude diagrams for the toroidal magnetic field, now covering both solar hemispheres, and exemplify the two generic classes of modulation arising in a nonlinear, non-kinematic dynamo model defined over a cartesian slab (see text). Figure kindly provided by S.M. Tobias, adapted from Beer et al. (1998).

coefficients, and it is not yet clear which aspects of the results are truly generic to Λ-quenching.

The dynamical backreaction of the large-scale magnetic field on meridional circulation has received comparatively little attention. The few calculations published so far[7] suggest that diffuse toroidal magnetic fields of strength up to 0.1 T can probably be advected equatorward at the core–envelope interface. That it can indeed do so is crucial for models relying on the meridional flow to produce equatorward propagation of magnetic fields as the cycle unfolds. Interestingly, relatively small variations of the meridional flow persisting over decadal timescales have been shown to produce cycle amplitude variations that are remarkably solar-like.[8] Such effects are potentially important in all dynamo models of the flux transport variety, and clearly merit further study.

---

[7] See paper by Rempel (2006) cited in the bibliography, and references therein.

[8] See paper by Lopes and Passos (2009) cited in the bibliography.

4.4 Cycle Modulation Through Time Delays

It was already noted that in solar cycle models based on the Babcock–Leighton mechanism of poloidal field generation, meridional circulation effectively sets—and even regulates—the cycle period. In doing so, it also introduces a long time delay in the dynamo mechanism, "long" in the sense of being comparable to the cycle period. This delay originates with the time required for circulation to advect the surface poloidal field down to the core–envelope interface, where the toroidal component is produced (A→C on Fig. 3.16). In contrast, the production of surface fields from the deep-seated toroidal component (C→D) is a "fast" process, growth rates and buoyant rise times for sunspot-forming toroidal flux ropes being of the order of tens of days. The first, long time delay turns out to have important dynamical consequences.

As proposed originally by B.R. Durney, the long time delay inherent in Babcock–Leighton models of the solar cycle allows a formulation of cycle-to-cycle amplitude variations in terms of a simple one-dimensional iterative map. Working in the kinematic regime, neglecting resistive dissipation, and in view of the conveyor belt argument outlined in Sect. 3.3, the toroidal field strength (T_{n+1}) at cycle $n+1$ is assumed to be linearly proportional to the poloidal field strength (P_n) of cycle n, i.e.,

$$T_{n+1} = a P_n \,. \tag{4.10}$$

Now, because flux eruption is a fast process, the strength of the poloidal field at cycle $n+1$ is (nonlinearly) proportional to the toroidal field strength of the *current* cycle:

$$P_{n+1} = f(T_{n+1}) T_{n+1} \,. \tag{4.11}$$

Here the "Babcock–Leighton" function $f(T_{n+1})$ measures the efficiency of surface poloidal field production from the deep-seated toroidal field. Substitution of Eq. (4.10) into Eq. (4.11) leads to a one-dimensional iterative map:

$$p_{n+1} = \alpha f(p_n) p_n \,, \qquad n = 0, 1, 2, \ldots \,, \tag{4.12}$$

where the p_n's are normalized amplitudes, and the normalization constants as well as the constant a in Eq. (4.10) have been absorbed into the definition of the map's parameter α, here operationally equivalent to a dynamo number. In analogy with Eq. (3.64), we adopt here the following nonlinear function

$$f(p) = \frac{1}{4}\left[1 + \mathrm{erf}\left(\frac{p-p_1}{w_1}\right)\right]\left[1 - \mathrm{erf}\left(\frac{p-p_2}{w_2}\right)\right] \,, \tag{4.13}$$

with $p_1 = 0.6$, $w_1 = 0.2$, $p_2 = 1.0$, and $w_2 = 0.8$. This catches an essential feature of the B–L mechanism, namely the fact that it can only operate in a finite range of toroidal field strength.

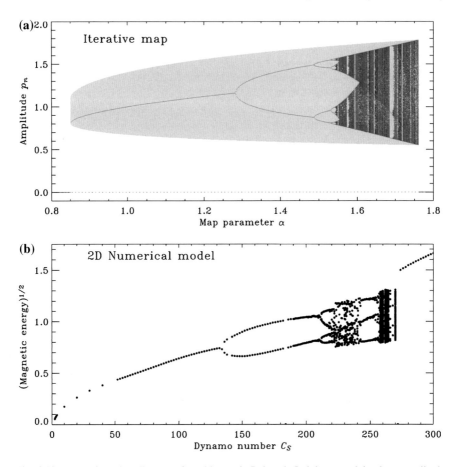

Fig. 4.10 Two bifurcation diagrams for a kinematic Babcock–Leighton model, where amplitude fluctuations are produced by time-delay feedback. The *top diagram* is computed using the one-dimensional iterative map given by Eqs. (4.12)–(4.13), while the *bottom diagram* is reconstructed from numerical 2D kinematic solutions in spherical geometry, of the type discussed in Sect. 3.3. The *shaded area* in panel **a** maps the attraction basin for the cyclic solutions, with initial conditions located outside of this basin converging to the trivial solution $p_n = 0$. From Living Review of Charbonneau (2010).

A bifurcation diagram for the resulting iterative map is presented on Fig. 4.10a. For a given value of the map parameter α, the diagram gives the locus of the amplitude iterate p_n for successive n values. The "critical dynamo number" above which dynamo action becomes possible, corresponds here to $\alpha = 0.851$ ($p_n = 0$ for smaller α values). For $0.851 \leq \alpha \leq 1.283$, the iterate is stable at some finite value of p_n, which grows gradually with α. This corresponds to a constant amplitude cycle. As α reaches 1.283, period doubling occurs, with the iterate p_n alternating between high and low values (e.g., $p_n = 0.93$ and $p_n = 1.41$ at $\alpha = 1.4$). Further period doubling occurs at $\alpha = 1.488$, then at $\alpha = 1.531$, then again at $\alpha = 1.541$, and ever faster until a point is reached beyond which the amplitude iterate seems to vary without any obvious pattern (although within a bounded range); this is in fact a chaotic regime.

As in any other dynamo model where the source regions for the poloidal and toroidal magnetic field components are spatially segregated, the type of time delay considered here is unavoidable. The Babcock–Leighton model is just a particularly clear-cut example of such a situation. One is then led to anticipate that the map's rich dynamical behavior should find its counterpart in the original, arguably more realistic spatially-extended, diffusive axisymmetric model that inspired the map formulation. Remarkably, this is indeed the case.

Figure 4.10b shows a bifurcation diagram, conceptually equivalent to that shown on part A, but now constructed from a sequence of numerical solutions of the Babcock–Leighton model discussed earlier in Sect. 3.3, for increasing values of the dynamo number. Time series of magnetic energy were calculated from the numerical solutions, successive peaks found and their peak amplitude plotted for each individual solution. The sequence of period doubling, leading to a chaotic regime, is strikingly similar to the bifurcation diagram constructed from the corresponding iterative map, down to the narrow multi-periodic windows interspersed in the chaotic domain. This demonstrates that time delay effects are a robust feature, and represent a very powerful source of cycle amplitude fluctuation in Babcock–Leighton models, *even in the kinematic regime*. Although transition to chaos does not always occur through such a classical period doubling sequence, chaos is ubiquitous in this model's parameter space.

4.5 Intermittency

The term "intermittency" refers to systems undergoing apparently random, rapid switching from quiescent to bursting behavior, as measured by the magnitude of some suitable system variable. In the context of solar cycle model, intermittency is invoked to explain the existence of Maunder Minimum-like quiescent epochs of strongly suppressed activity, randomly interspersed within periods of "normal" cyclic activity.[9]

Intermittency has been shown to occur through stochastic fluctuations of the dynamo number in mean-field dynamo models operating at or near criticality.[10] This mechanism for "on-off intermittency" works well, however there is no strong

---

[9] It should be noted, however, that dearth of sunspots does not necessarily mean a halted cycle; as noted earlier, flux ropes of strengths inferior to ~ 1 T will not survive their rise through the convective envelope, and the process of flux rope formation from the dynamo-generated mean magnetic field may itself be subjected to a threshold in field strength. The same basic magnetic cycle may well have continued unabated all the way through the Maunder Minimum, but at an amplitude just below one of these thresholds. This idea finds support in the ^{10}Be radioisotope record, which shows an uninterrupted cyclic signal through the Maunder minimum (see Fig. 4.5).

[10] See paper by Ossendrijver & Hoyng (1996) cited in the bibliography for a particularly lucid discussion.

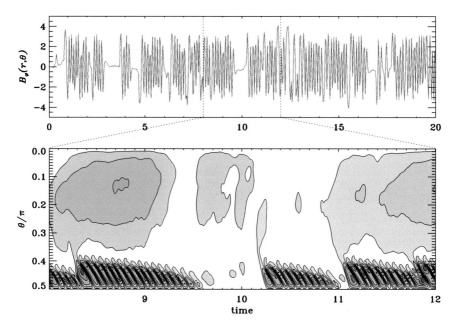

Fig. 4.11 Intermittency in a dynamo model based on flux tube instabilities (cf. Sect. 3.4.2). The *top panel* shows a trace of the toroidal field, and the *bottom panel* is a butterfly diagram covering a shorter time span including a quiescent phase at $9.6 \lesssim t \lesssim 10.2$, and a "failed Minimum" at $t \simeq 11$. Figure produced from numerical data kindly provided by M. Ossendrijver. From Living Review of Charbonneau (2010).

reason to believe that the solar dynamo is running just at criticality, so that it is not clear how good an explanation this is of Maunder-type grand minima. Parity modulation driven by stochastic noise can also lead to a form of intermittency in linear or α-quenched models, by exciting the higher-order modes that perturb the normal operation of the otherwise dominant dynamo mode, producing marked hemispheric asymmetries and strongly reduced cycle amplitudes. The transition from active to quiescent (and vice versa) being controlled by stochastic noise, the durations of active and quiescent phases tend to have exponential distributions, which agrees tolerably well with inferences from the radioisotope records.[11]

Another way to trigger intermittency in a dynamo model is to let nonlinear dynamical effects, for example a reduction of the differential rotation amplitude, push the effective dynamo number below its critical value; dynamo action then ceases during the subsequent time interval needed to reestablish differential rotation following the diffusive decay of the magnetic field; in the low P_m regime, this time interval can amount to many cycle periods, but P_m must not be too small, otherwise grand minima

---

[11] For more on this version of noise-driven intermittency, see the papers by Mininni and Gómez (2002) and Moss et al. (2008) cited in the bibliography.

4.5 Intermittency

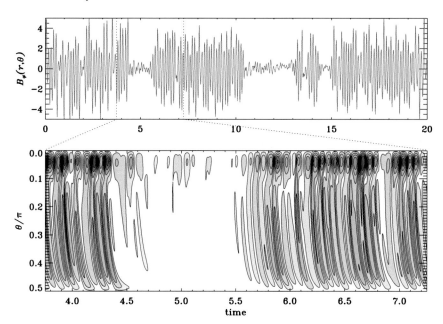

Fig. 4.12 Intermittency in a dynamo model based on the Babcock–Leighton mechanism (cf. Sect. 3.3). The *top panel* shows a trace of the toroidal field sampled at $(r, \theta) = (0.7, \pi/3)$. The *bottom panel* is a time-latitude diagram for the toroidal field at the core–envelope interface. From Living Review of Charbonneau (2010).

become too rare. Values $P_m \sim 10^{-2}$ seems to work best. Such intermittency is again most readily produced when the dynamo is operating close to criticality.[12]

Intermittency has also been observed in strongly supercritical models including α-quenching as the sole amplitude-limiting nonlinearity. Such solutions can enter grand minima-like epochs of reduced activity when the dynamo-generated magnetic field completely quenches the α-effect. The dynamo cycle restarts when the magnetic field resistively decays back to the level where the α-effect becomes operational once again.[13]

Intermittency can also arise naturally in dynamo models characterized by a lower operating threshold on the magnetic field. These include models where the regeneration of the poloidal field takes place via the MHD instability of toroidal flux tubes (Sect. 3.4.2). In such models, the transition from quiescent to active phases requires an external mechanism to push the field strength back above threshold. This can be stochastic noise,[14] or a secondary dynamo process normally overpowered by the "primary" dynamo during active phases.

---

[12] See, e.g., the papers by Küker et al. (1999) and Brooke et al. (2002) cited in the bibliography.

[13] For representative models exhibiting intermittency of this type, see the paper by Tworkowski et al. (1998) cited in the bibliography.

[14] See the paper by Schmitt et al. (1996) cited in the bibliography.

Figure 4.11 show one representative solution of the latter variety, where intermittency is driven by a weak α-effect-based kinematic dynamo operating in the convective envelope, in conjunction with magnetic flux injection into the underlying region of primary dynamo action by randomly positioned downflows.[15] The top panel shows a sample trace of the toroidal field, and the bottom panel a butterfly diagram constructed near the core–envelope interface in the model.

Dynamo models exhibiting amplitude modulation through time-delay effects are also liable to intermittency in the presence of stochastic noise. This intermittency mechanism hinges on the fact that the map's attractor has a finite basin of attraction (indicated by gray shading on Fig. 4.10a). Stochastic noise acting in conjunction with the map's dynamics can then knock the solution out of this basin of attraction, which then leads to a collapse onto the trivial solution $p_n = 0$, even if the map parameter remains supercritical. Stochastic noise eventually knocks the solution back into the attractor's basin, which signals the onset of a new active phase. This behavior does materialize in the Babcock–Leighton model of Sect. 3.3. Figure 4.12 shows one such representative solution, in the same format as Fig. 4.11. This is a dynamo solution which, in the absence of noise, operates in the singly-periodic regime. Stochastic noise is added to the vector potential $A\hat{e}_\phi$ in the surface layers, and the dynamo number is also allowed to fluctuate randomly about a pre-set mean value, as described in Sect. 4.2. The resulting solution exhibits both amplitude fluctuations and intermittency.

With its strong polar branch often characteristic of dynamo models with meridional circulation, Fig. 4.12 is not a particularly good fit to the sunspot butterfly diagram. Yet its fluctuating behavior is solar-like in a number of ways, including epochs of alternating higher-than-average and lower-than-average cycle amplitudes (the Gnevyshev–Ohl rule, cf. Fig. 4.3), and residual pseudo-cyclic variations during quiescent phases, as suggested by ^{10}Be data. This latter property is due at least in part to meridional circulation, which continues to advect the (diffusively decaying) magnetic field after the dynamo has fallen below threshold.

4.6 Model-Based Cycle Predictions

Over the past decade, the prediction of solar eruptive events and their geomagnetic impacts, known as *space weather*, has become a Very Big Business. Even then, the prediction of the overall level of solar activity is also of interest, as it could be useful, among other things, to the planning of space missions and interplanetary travel. The understanding and prediction of activity levels on timescales decadal and longer is becoming known as *space climate*, and its primary data are the time series of sunspot numbers, and proxies such as the radioisotopes records.

One "hot" prediction problem, lying at the boundary of space weather and space climate, is forecasting the characteristic of the next solar activity cycle, which is usually equated with the timing and amplitude of the cycle as measured by the sunspot

---

[15] For more details see paper by Ossendrijver (2000) in the bibliography.

4.6 Model-Based Cycle Predictions

number time series (see Fig. 4.1). It is of course possible to treat this prediction problem as an exercise in time series analysis and forecasting, without any physical input. The SSN time series is just a time series, and it can be extended using a number of techniques coming from statistics (spectral analysis, wavelets, etc.) or dynamical system theory (such as attractor reconstruction). To this day, forecasts based on such techniques have not fared significantly better than so-called "climatological" forecasts, which consists in simply "predicting", e.g., that the next cycle will have the same amplitude as the current cycle, or an amplitude equal to the mean cycle amplitude over the length of the sunspot record, etc. In this section, we will focus instead on prediction schemes based, in one form or another, on dynamo models.

In light of what we have learned thus far, we know we are facing a number of difficulties in trying to use dynamo models to forecast the solar cycle. A basic list of questions that need to be answered (excluding technical details for the time being) should include, at the very least, the following:

1. What type of dynamo powers the solar cycle: $\alpha\Omega$, $\alpha^2\Omega$, interface, Babcock–Leighton, etc.?
2. Which mechanism is driving duration and amplitude fluctuations: stochastic forcing, nonlinear modulation due to the Lorentz force, or time delay, etc.?
3. How do we accurately "predict" sunspot number from a dynamo solution which describes the spatiotemporal evolution of just the diffuse, large-scale magnetic field?

It is a sobering fact that none of these very basic and fundamental questions can be answered with confidence at this writing. Nonetheless, we have learned some important things that are useful in the forecasting context. To start with, the dynamo feeds on the existing magnetic field, therefore trying to forecast the next cycle using characteristics of the current cycle (and maybe recent past cycles as well) is definitely justified. This is the physical underpinning of all so-called "precursor methods", which we'll first look into.

4.6.1 The Solar Polar Magnetic Field as a Precursor

Temporally extended synoptic magnetograms, such as Fig. 3.4, suggest that the solar cycle can be divided into sequences of substeps whereby a poloidal field (P) produces a new toroidal component (T), which then leads to the buildup of a new poloidal component, with accompanying polarity reversals; schematically:

$$\cdots \to P(+) \to T(-) \to P(-) \to T(+) \to P(+) \to \cdots . \qquad (4.14)$$

This suggests that the optimal precursor for the amplitude of the sunspot-generating toroidal component should be sought by moving back up the causal chain by one substep, to the poloidal component produced in the previous sunspot cycle. This is the basis for the set of dynamo-inspired precursor schemes pioneered by A.L. Ohl

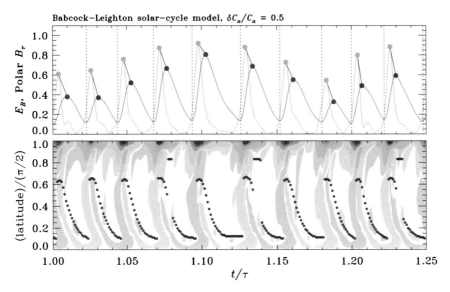

Fig. 4.13 Portion of a simulation run of a Babcock–Leighton model, with fluctuations at the ±50% level imposed in the magnitude of the surface source term. The unperturbed reference solution is that illustrated on Fig. 3.17. The *top panel* shows time series of the surface radial magnetic field (absolute value) sampled at the pole (*green*), together with a time series of the total magnetic energy (*red*), used here as a proxy for the sunspot number. The purple line segment join the peak poloidal field at (or near) "sunspot minimum" with the peak in the SSN proxy for the following cycle. The *bottom panel* is a time–latitude diagram of the surface radial field, and the purple dots trace the latitude of peak toroidal field strength at the core–envelope interface as a function of time. Figure adapted from the Charbonneau & Barlet (2011) paper cited in the bibliography with permission from Elsevier.

and brought to maturity by K. Schatten et al. (1978) now over thirty years ago (see bibliography).

This idea is readily tested using our various dynamo models, as illustrated on Fig. 4.13 for the stochastically forced Babcock–Leighton model of Fig. 4.8. The top panel shows a short segment of the magnetic energy time series, used again here as a proxy for the sunspot number, together with a time series of the surface polar field strength (in green). The bottom panel shows a time–latitude diagram of the surface radial magnetic component, together with the latitudes of peak toroidal field strength at the core–envelope interface, where sunspots are presumed to originate. The overall spatiotemporal evolution of the surface field, and its phase relationship to the deep-seated toroidal field, are both remarkably solar-like. Examination of the two curves on the top panels of Fig. 4.13 reveals that the surface radial field peaks shortly following what one would identify with solar minimum on the basis of our SSN proxy. It is then a simple matter to pair the peak polar field at solar minimum with the SSN proxy of the following cycle, as indicated on Fig. 4.13 by the purple connecting line segments.

4.6 Model-Based Cycle Predictions

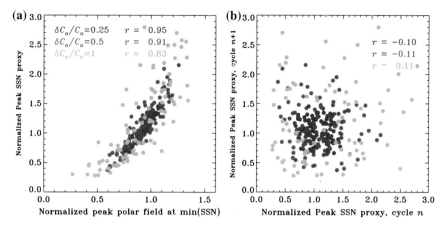

Fig. 4.14 a Correlations between peak SSN and surface poloidal field strength in the stochastically-forced Babcock–Leighton solutions of Fig. 3.17, for three different levels $\delta C_\alpha / \bar{C}_\alpha$ of forced stochastic fluctuations in the surface source term, color-coded as indicated. The *red dots* correspond to the simulation run illustrated on Fig. 4.13. The amplitudes are normalized to those characterizing the non-fluctuating parent simulation. The linear correlation coefficients r are again given. **b** Similar plot, but this time attempting to correlate the SSN peak values for pairs of successive cycles. Adapted from Charbonneau & Barlet (2011) with permission from Elsevier.

The next step is to correlate the peak poloidal field and peak SSN proxy, in order to ascertain the viability of poloidal field-based precursor schemes. The result is shown on Fig. 4.14a, for three different levels of stochastic forcing, as color-coded, with the solution of Fig. 4.13 in red. In all cases the time series for the SSN proxies and surface poloidal field strength have been normalized to the peak values characterizing a parent run without stochastic forcing.

The peak polar field at solar minimum clearly has precursor value, but stochastic forcing rapidly degrades the forecasting accuracy. Consider for example the solution with 100% fluctuation of the dynamo number C_α (green); while a linear correlation coefficient of 0.83 may sound pretty good, the fact remains that for a polar field of 0.8 (say) in the normalized units of Fig. 4.14 would lead to a SSN forecast covering a very broad range, namely 0.6–1.2 in normalized units, which is not a very accurate forecast at all.

Performing the same analysis on our other solar cycle model reveals that the polar surface field is also a good precursor of cycle amplitude for the advection-dominated mean-field model with meridional circulation of Sect. 3.2.11, more robust with respect to high-amplitude stochastic forcing in fact, but no precursor at all for the classical $\alpha\Omega$ model of Sect. 3.2.10. This curious situation can be traced to the fact that in the former, the surface polar field does feed back into the dynamo loop, as circulation drags it down back into the tachocline, where it merges with the poloidal field produced there by the α-effect (see Fig. 3.10). In the circulation-free models, on the other hand, the poloidal field diffuses more or less radially outwards to the surface, with poloidal field of the subsequent cycle being generated completely independently at the base of the envelope (see Fig. 3.7).

In retrospect, the logic behind Schatten et al.'s precursor argument can be understood to hold only for a subset of dynamo models, namely those where some "feedback" of the surface polar field on the dynamo loop takes place. In Babcock–Leighton-type models, the surface field is the sole source of the next cycle's toroidal field, and so is a good precursor. In advection-dominated mean-field models including circulation, the surface poloidal field is a significant source of toroidal field, albeit not the only one. In classical $\alpha\Omega$ mean-field models of the type considered in Sect. 3.2.10, where the surface field is only a "passive" manifestation of dynamo action taking place independently in the deep interior, the surface field has no precursor value.

As final point of interest, it is noteworthy that in the solar cycle models considered here, the value for the peak SSN proxy has little or no precursor value in forecasting the next SSN proxy peak. This is illustrated on Fig. 4.14b for the set of stochastically-forced Babcock–Leighton solutions. This is somewhat surprising, since the peak polar field at the solar minimum separating two successive cycles is here a rather good precursor (cf. Fig. 4.14a). This situation can be traced to the manner in which stochasticity is introduced in the model. In the case of imposed stochastic fluctuations in the poloidal source term, the scheme given by Eq. (4.14) must be replaced by something like:

$$\cdots \xrightarrow{\text{stoch}} P(+) \longrightarrow T(-) \xrightarrow{\text{stoch}} P(-) \longrightarrow T(+) \xrightarrow{\text{stoch}} P(+) \longrightarrow \cdots. \quad (4.15)$$

Precursor forecasts based on either component is only possible if the forecast does not go across a "stochastic" arrow. For classical mean-field models, where the shear and α-effect are spatially coincident and operate concurrently in time, the above sequence should instead be schematized as:

$$\cdots \overset{\text{stoch}}{\underset{T(-)}{\cancel{}}} \begin{matrix}P(+)\\ \end{matrix} \overset{\text{stoch}}{\underset{T(+)}{\cancel{}}} \begin{matrix}P(-)\\ \end{matrix} \overset{\text{stoch}}{\underset{T(-)}{\cancel{}}} \begin{matrix}P(+)\\ \end{matrix} \overset{\text{stoch}}{\cancel{}} \cdots, \quad (4.16)$$

which precludes any precursor relationship from previous-cycle magnetic field measurements …unless of course stochastic forcing is very weak or absent, and cycle amplitude modulation is produced primarily by deterministic effects, as briefly considered in Sect. 4.3.

4.6.2 Model-Based Prediction Using Solar Data

Some recent solar cycle amplitude forecasts have used solar cycle models of the Babcock–Leighton variety (Sect. 3.3), in conjunction with input of solar magnetic field observations in a manner usually (and, strictly speaking, incorrectly) described

4.6 Model-Based Cycle Predictions

Table 4.1 Two dynamo-based solar cycle forecasting schemes

Authors	Dikpati, deToma & Gilman	Choudhuri, Chatterjee & Jiang
Code name	DdTG	CCJ
Reference	GRL **33**, L05102 (2006)	PRL **98**, 131103 (2007)
Dynamo model	Kinematic axisymmetric Babcock–Leighton	Kinematic axisymmetric Babcock–Leighton
Core–CZ interface	$r/R = 0.7$	$r/R = 0.7$
Magnetic diffusivity	Eq. (2.16), $\Delta\eta = 300$ plus high-η surface layer	Eq. (2.16), $\Delta\eta = 10^4$
Differential rotation	Solar-like parameterization Eqs. (2.27) and (2.28), $w/R = 0.05$ all other parameters same	Solar-like parameterization Eqs. (2.27) and (2.28), $w/R = 0.015$ all other parameters same
Meridional circulation	Single cell per quadrant closing at $r/R = 0.71$	Single cell per quadrant closing at $r/R = 0.635$
Poloidal source term	Data-driven surface forcing plus weak tachocline α-effect	Subsurface α-effect plus buoyancy algorithm
Nonlinearity	Algebraic α-quenching only in tachocline α-effect	Algebraic α-quenching in subsurface α-effect
Solar data	Time series of total sunspot area used to (continuously) drive parametric surface forcing of A	DM Index used to reset amplitude of A at "solar minimum"
Calibration interval	Cycles 16–23	Cycles 21–23
Cycle 24 forecast	SSN = 155–180	SSN = 80

as "data assimilation". It is particularly instructive to compare the forecast schemes (and cycle 24 predictions) of M. Dikpati and collaborators on the one hand, and of A.R. Choudhuri and collaborators on the other. As detailed in Table 4.1 below, these two schemes are remarkably similar in their overall design, differing mostly in their formulation of the poloidal source term, solar data used to drive the model, and manner in which this driving is implemented.

As similar as they may be, except at the level of what one would usually consider modelling details, these two forecasting schemes end up producing cycle 24 amplitude forecasts that stand at opposite ends of the very wide range of cycle 24 forecasts produced by other techniques, as well as opposite ends of the range of past cycle amplitudes. A cycle 24 with SSN = 80 would place it amongst the weakest of the past century, while SSN = 180 would make it second only to the highest cycle amplitude on record, that for cycle 19 (see Fig. 4.1).

These model-based forecasts have been subjected to strident criticism, for a variety of reasons. One of the most fundamental is the possibility—some would say "certainty"—that the solar dynamo is a nonlinear system operating in the chaotic regime, in which case long-term prediction is severely restricted by the exponential divergence of trajectories of the model in phase space. This criticism probably does not apply to the DdTG scheme, which is really a quasi-linear magnetic flux processing "machine", rather than a truly nonlinear dynamo model; it probably does not apply either to the CCJ scheme, which uses a simple algebraic amplitude-quenching nonlinearity that is usually not conducive to chaotic modulation, although this remains to be verified in the context of their specific choice of dynamo model. More to the point has been the explicit demonstration that very small changes in some unobservable and poorly constrained input parameters to the dynamo model used for the forecast, or alternate but by all appearances equally reasonable means of carrying out data input into the model, can introduce significant errors already for next-cycle amplitude forecasts.[16]

In the context of Babcock–Leighton models, this model-based approach to forecasting is definitely viable in principle, since the solar surface magnetic field is that which will serve as seed to produce the sunspot-generating toroidal component of the next cycle. The one thing that the two model-based forecasting schemes compared and contrasted in Table 4.1 have demonstrated, beyond any doubt, is that modelling details matter a lot.

Bibliography

> The possible impact of long-term variations of solar activity on climate change remains a topic of controversy; the following three volumes are a good starting point for those interested in learning more about this:

Haigh, J. D., Lockwood, M., & Giampapa, M. S.: 2005, Saas-Fee Advanced Course, Vol. 34, *The Sun, Solar Analogs and the Climate*, ed. I. Rüedi, M. Güdel, & W. Schmutz, Springer

Benestad, R. E.: 2006, *Solar Activity and Earth's Climate*, 2nd edn., Springer

Schrijver, C. J., & Siscoe, G. L.: 2010, *Heliophysics III: Evolving Solar Activity and the Climates of Space and Earth*, Cambridge University Press

> On pre- and early-telescopic observations of sunspots, begin with

Mitchell, W. M.: 1916, *The history of the discovery of the solar spots*, Popular Astronomy, **24**, 22-ff

Vaquero, J. M., & Vázquez, M.: 2009, Astrophysics and Space Science Library, Vol. 361, *The Sun Recorded Through History: Scientific Data Extracted from Historical Documents*, Springer

Reeves, E., & van Helden, A., eds.: 2010, *Galileo Galilei & Christoph Scheiner on Sunspots*, University of Chicago Press

> If such historical matters are of interest to you, you can also consult the ever-being-enlarged Web site "Great Moments in the History of Solar Physics":

http://www.astro.umontreal.ca/~paulchar/grps

> and click on "History of Solar Physics" at left. For everything you ever wanted to know on the characterisations of the sunspot cycle, start with

Hathaway, D. H.: 2010, *The solar cycle*, Liv. Rev. Solar Phys., **7**, 1, http://solarphysics.livingreviews.org/Articles/lrsp-2010-1/

---

[16] See, e.g., the papers by Bushby & Tobias (2007) and Cameron & Schüssler (2007) cited in the bibliography.

Bibliography

On the Maunder minimum, see

Eddy, J. A.: 1976, *The Maunder minimum*, Science, **192**, 1189–1202
Eddy, J. A.: 1983 *The Maunder minimum - A reappraisal*, Solar Phys. **89**, 195–207
Ribes, J. C., & Nesme-Ribes, E.: 1993 *The solar sunspot cycle in the Maunder minimum AD1645 to AD1715*, Astron. & Astrophys., **276**, 549–563

and on cosmogenic radioisotopes:

Beer, J.: 2000, *Long-term indirect indices of solar variability*, Space Sci. Rev., **94**, 53–66
Usoskin, I. G., Solanki, S. K., & Kovaltsov, G. A.: 2007, *Grand minima and maxima of solar activity: new observational constraints*, Astron. & Astrophys., **471**, 301–309

as well as the chapter by J. Beer in the volume edited by Schrijver & Siscoe just listed above. On the effects of stochastic forcing on various dynamo models, start with:

Choudhuri, A. R.: 1992, *Stochastic fluctuations of the solar dynamo*, Astron. & Astrophys., **253**, 277–285
Moss, D., Brandenburg, A., Tavakol, R., & Tuominen, I.: 1992, *Stochastic effects in mean-field dynamos*, Astron. & Astrophys., **265**, 843–849
Hoyng, P.: 1993, *Helicity fluctuations in mean field theory: an explanation for the variability of the solar cycle?*, Astron. & Astrophys., **272**, 321–339
Ossendrijver, M. A. J. H., & Hoyng, P.: 1996, *Stochastic and nonlinear fluctuations in a mean field dynamo*, Astron. & Astrophys., **313**, 959–970
Charbonneau, P., & Dikpati, M.: 2000, *Stochastic fluctuations in a Babcock-Leighton model of the solar cycle*, Astrophys. J., **543**, 1027–1043
Mininni, P. D., & Gómez, D. O.: 2002, *Study of stochastic fluctuations in a shell dynamo*, Astrophys. J., **573**, 454–463

The following offer a good sample of the possible amplitude and parity modulation behaviors in nonlinear (sometimes non-kinematic) mean-field dynamo models:

Brooke, J., Moss, D., & Phillips, A.: 2002, *Deep minima in stellar dynamos*, Astron. & Astrophys., **395**, 1013–1022
Küker, M., Arlt, R., & Rüdiger, G.: 1999, *The Maunder minimum as due to magnetic Λ-quenching*, Astron. & Astrophys., **343**, 977–982
Sokoloff, D., Nesme-Ribes, E.: 1994, *The Maunder minimum: A mixed-parity dynamo mode?*, Astron. & Astrophys., **288**, 293–298
Tobias, S. M.: 1997, *The solar cycle: parity interactions and amplitude modulation*, Astron. & Astrophys., **322**, 1007–1017
Bushby, P. J.: 2006, *Zonal flows and grand minima in a solar dynamo model*, Mon. Not. Roy. Astron. Soc., **371**, 772–780
Rempel, M.: 2006, *Flux-transport dynamos with Lorentz-force feedback on differential rotation and meridional flow: saturation mechanism and torsional oscillations*, Astrophys. J., **647**, 662–675
Lopes, I., & Passos, D.: 2009, *Solar variability induced in a dynamo code by realistic meridional circulation variations*, Solar Phys., **257**, 1–12

Figure 4.9 is an adaption of Fig. 6 of

Beer, J., Tobias, S., & Weiss, N.: 1998, *An active sun throughout the Maunder minimum*, Solar Phys., **181**, 237–249

On time delay and its consequences for Babcock-Leighton dynamo models, see:

Durney, B. R.: 2000, *On the differences between odd and even solar cycles*, Solar Phys., **196**, 421–426
Charbonneau, P.: 2001, *Multi-periodicity, chaos, and intermittency in a reduced model of the solar cycle*, Solar Phys., **199**, 385–404

Charbonneau, P., St-Jean, C., & Zacharias, P.: 2005, *Fluctuations in Babcock-Leighton dynamos. I. Period doubling and transition to chaos*, Astrophys. J., **619**, 613–622

The following offers a few good entry points in the literature on intermittency in various types of dynamo models:

Schmitt, D., Schüssler, M., & Ferriz-Mas, A.: 1996, *Intermittent solar activity by an on-off dynamo*, Astron. & Astrophys., **311**, L1–L4

Tworkowski, A., Tavakol, R., Brandenburg, A., Brooke, J. M., Moss, D., & Tuominen, I.: 1998 *Intermittent behaviour in axisymmetric mean-field dynamo models in spherical shells*, Mon. Not. Roy. Astron. Soc., **296**, 287–295

Covas, E., & Tavakol, R.: 1999, *Multiple forms of intermittency in partial differential equation dynamo models*, Phys. Rev. E, **60**, 5435–5438

Ossendrijver, M. A. J. H.: 2000, *Grand minima in a buoyancy-driven solar dynamo*, Astron. & Astrophys., **359**, 364–372

Charbonneau, P., Blais-Laurier, G., & St-Jean, C.: 2004, *Intermittency and phase persistence in a Babcock-Leighton model of the solar cycle*, Astrophys. J. Lett., **616**, L183–L186

Petrovay, K.: 2007, *On the possibility of a bimodal solar dynamo*, Astron. Nachr., **328**, 777–780

Moss, D., Sokoloff, D., Usoskin, I., & Tutubalin, V.: 2008, *Solar grand minima and random fluctuations in dynamo parameters*, Solar Phys., **250**, 221–234

Figures 4.10 and 4.11 are from

Charbonneau, P.: 2010, *Dynamo models of the solar cycle*, Liv. Rev. Solar Phys., **7**, 3. http://solarphysics.livingreviews.org/Articles/lrsp-2010-3/

Cycle prediction is a topic that has generated a massive literature, which often stands closer to statistical black magic than physics. Two good recent review papers are:

Hathaway, D. H.: 2009, *Solar cycle forecasting*, Space Sci. Rev. **144**, 401–412

Petrovay, K.: 2010, *Solar cycle prediction*, Liv. Rev. Solar Phys., **7**, 6. http://solarphysics.livingreviews.org/Articles/lrsp-2010-6/

On dynamo-inspired precursor schemes, see

Schatten, K. H., Scherrer, P. H., Svalgaard, L., & Wilcox, J. M.: 1978, *Using dynamo theory to predict the sunspot number during solar cycle 21*, Geophys. Res. Lett., **5**, 411–414

Svalgaard, L., Cliver, E. W., & Kamide, Y.: 2005, *Sunspot cycle 24: smallest cycle in 100 years?*, Geophys. Res. Lett., **32**, id. L01104

Charbonneau P, & Barlet, G.: 2011, *The dynamo basis of solar cycle precursor schemes*, J. Atmos. Solar-Terrestrial Phys., **73**, 198–206

On the use of dynamo models for cycle prediction, see

Dikpati, M., de Toma, G., & Gilman, P. A.: 2006, *Predicting the strength of solar cycle 24 using a flux-transport dynamo-based tool*, Geophys. Res. Lett., **33**, id. L05102

Choudhuri, A. R., Chatterjee, P., & Jiang, J.: 2007, *Predicting solar cycle 24 with a solar dynamo model*, Phys. Rev. Lett., **98**, id. 131103

Cameron, R., & Schüssler, M.: 2007, *Solar cycle prediction using precursors and flux transport models*, Astrophys. J., **659**, 801–811

as well as

Bushby, P. J., & Tobias, S. M.: 2007, *On predicting the solar cycle using mean-field models*, Astrophys. J., **661**, 1289–1296

This last paper is an illuminating discussion of the fundamental limitations inherent in using nonlinear dynamo models for cycle amplitude forecasting. Keep in mind, however, that the two forecasting models discussed in Sect. 4.6.2 are (1) not operating in a chaotic regime, and (2) achieve their forecasting not just through direct forward integration, but also through continuous or episodic input of observational data. Finally, the following paper is a very interesting example of true data assimilation in a very simple albeit truly nonlinear dynamo model:

Kitiashvili, I., & Kosovichev, A. G.: 2008, *Application of data assimilation method for predicting solar cycles*, Astrophys. J. Lett., **688**, L49–L52

Chapter 5
Stellar Dynamos

> *ELWOOD: It's 106 miles to Chicago, we've got a full tank of gas, half a pack of cigarettes, it's dark and we're wearing sunglasses. JAKE: Hit it!*
>
> Dan Ackroyd and John Belushi
> The Blues Brothers (1980)

The problem—and the beauty—with the sun is that it overwhelms us with data. Many of the intricacies we have busied ourselves with in the preceding chapter were directly motivated by the detailed observations and magnetic field measurements made possible by the sun's astronomical proximity. The sun remains for sure an exemplar, but with other stars, observational constraints are much more sparse, and theoretical considerations take on an enlarged role.

What have we learned in the preceding three chapters about dynamo action in electrically conducting fluids? At the most fundamental level, a top-three list could run as follows:

- We learned in Chap. 2 that rotation, and especially differential rotation, is one very powerful mechanism allowing to build up a large-scale magnetic field;
- We also learned in Chap. 2 that flows with chaotic trajectories, such as arising from turbulent convection, can act as dynamos;
- We learned in Chap. 3 that in turbulent flows, the presence of rotation and stratification can break isotropy and reflectional symmetry, and in doing so generate a mean electromotive force that can produce large-scale magnetic fields.

So, offhand we are not in too bad a shape with regards to stellar dynamos. Stars certainly are stratified, and certainly rotate. Thermally-driven convection is also present across large parts of the HR diagram, but here we start to encounter complications that restrict the use of the "solar exemplar".

Figure 5.1 illustrates, in schematic form, the internal structure of main-sequence stars, more specifically the presence or absence of convection zones. A G-star like the sun has a thick outer convection zone, spanning the outer 30 % in radius in the

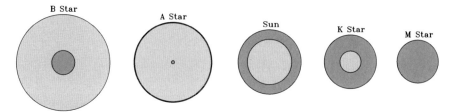

Fig. 5.1 Schematic representation of the radiative/convective internal structure of main-sequence stars. The thickness of the outer convection zone for the A-star is here greatly exaggerated; drawn to scale it would be thinner than the *black circle* delineating the stellar surface on this drawing. Relative stellar sizes are also not to scale; a main-sequence M0 star has a radius some 12 times smaller than its B0 counterpart.

solar case. As one moves down to less massive stars, the relative thickness of the convective envelope increases until, somewhere around spectral type M5, stars become fully convective. Moving instead from the sun to higher masses, the convective envelope becomes ever thinner, until somewhere around spectral type A0 it essentially vanishes. However, at around the same spectral type Hydrogen burning switches from the pp chain to the CNO cycle, for which nuclear reaction rates are much more sensitively dependent on temperature. Core energy release becomes strongly depth-dependent, leading to a steep—and convectively unstable—temperature gradient. This produces a small convective core, which grows in size as one moves up to larger masses. In a "typical" B-star of solar metallicity, the convective core spans the inner 25% or so in radius of the star.

In main-sequence O and B stars, the presence of a turbulent convective core combined with high rotation then makes dynamo action more than likely. However, as we shall see in Sect. 5.1 below, the challenge is actually to bring the magnetic field produced in the core to the surface.

Intermediate mass main-sequence stars rotate but lack a convective zone of substantial size. Interestingly, only the most slowly rotating of these stars show evidence of strong magnetic fields, for which the fossil hypothesis remains the favored explanation. This is reviewed in Sect. 5.2, along with some dynamo-based alternatives.

Until strong evidence to the contrary is brought to the fore, we are allowed to assume that late-type stars with a thick convective envelopes overlying a radiative core host a solar-type dynamo. This is buttressed by the observation of solar-like cyclic activity in many such stars. It then becomes natural to look into the way(s) the various types of solar-cycle models considered in the preceding chapters can be "scaled" to other solar-type stars, of varying masses, rotation rates, etc. Some of these issues are discussed in Sect. 5.3 below.

With fully convective stars (Sect. 5.4), we encounter potential deviations from a solar-type dynamo mechanism; without a tachocline and radiative core to store and amplify toroidal flux ropes, the Babcock–Leighton mechanism becomes problematic. Mean-field models based on the turbulent α-effect remain viable, but the dynamo behavior becomes dependent on the presence and strength of differential rotation, about which we really don't know very much in stars other than the sun.

Moving off the main-sequence, stellar magnetic fields remain ubiquitous, but dynamo modelling is comparatively undeveloped. Observational evidence points towards turbulent dynamo action in pre-main-sequence stars, as briefly reviewed in Sect. 5.5. On the other hand, the very strong magnetic fields observed in the many types of compact objects marking the endpoint of stellar evolution appear, for the most part, to be remnants of earlier evolutionary phases, rather than being produced by ongoing dynamo action (Sect. 5.6).

5.1 Early-Type Stars

5.1.1 Mean-Field Models

We first consider dynamo action in massive stars, beginning with a few simple, representative solutions obtained in the framework of mean-field theory.[1] Within the convective core (radius r_c), thermally-driven turbulent fluid motions are assumed to give rise to an α-effect and turbulent diffusivity, which both vanish for $r \gtrsim r_c$ (under the assumption that the radiative envelope is turbulence free). In the spirit of the other mean-field models discussed earlier, we consider kinematic dynamos with parametric profiles for α and η:

$$\alpha(r, \theta) = \frac{1}{2}\left[1 + \mathrm{erf}\left(\frac{r - r_c}{w}\right)\right] \mathrm{erf}\left(\frac{2r}{w}\right) \cos(\theta), \tag{5.1}$$

$$\eta(r) = \eta_e + \frac{\eta_c - \eta_e}{2}\left[1 - \mathrm{erf}\left(\frac{r - r_c}{w}\right)\right], \tag{5.2}$$

where $\mathrm{erf}(x)$ is the error function. Note that now, unlike in the solar models of the preceding chapters, it is the low-diffusivity stable envelope that sits atop the high-diffusivity convective core, i.e., now $\eta_e \ll \eta_c$. Equation (5.1) once again represent "minimal" assumptions on the spatial dependency of the α-effect: it changes sign across the equator ($\theta = \pi/2$), vanishes at $r = 0$, rises to a maximum value within the convective core, and falls again to zero for $r \gtrsim r_c$, the transition occurring across a spherical layer of thickness $\sim 2w$. Little being known about the turbulent structure of convective cores in massive stars, in what follows even the sign of the α-effect will be treated as a free parameter.

All dynamo solutions discussed below are obtained as eigenvalue problems, as in Sect. 3.2.8, solving now the α^2 form of the axisymmetric kinematic mean-field dynamo equations (Eqs. (3.33)–(3.34)). Remember that such linear solutions leave the absolute scale of the magnetic field unspecified. However, an interesting physical

---

[1] The content of this section is based on the paper by Charbonneau and MacGregor (2001) given in the bibliography.

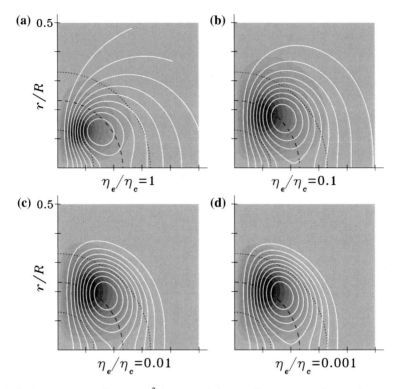

Fig. 5.2 Four antisymmetric steady α^2 dynamo solutions (**a–d**), computed using varying magnetic diffusivity ratios between the core and envelope. The solutions are plotted in a meridional quadrant, with the symmetry axis coinciding with the *left* quadrant boundary. Poloidal field lines are plotted superimposed on a *gray scale* representation for the toroidal field (*light* to *dark* is weaker to stronger field). The *dashed line* marks the core–envelope interface depth r_c, and the two *dotted lines* indicates the depths $r_c \pm w$ corresponding to the width of the transition layer between core and envelope. These solutions have a surface-to-core magnetic field strength ratio $\Sigma \simeq 10^{-2}$ at $\eta_e/\eta_c = 1$, down already to 3×10^{-4} at $\eta_e/\eta_c = 10^{-1}$ and falling below 10^{-8} for $\eta_e/\eta_c \lesssim 10^{-2}$. Reproduced from Charbonneau and MacGregor (2001) by permission of the AAS.

quantity accessible from linear models is the ratio of the surface field strength to the field strength in the dynamo region, here the convective core. In what follows we use towards this purpose the ratio (Σ) of the r.m.s. surface poloidal field to the r.m.s. poloidal field at the core–envelope interface r_c:

$$\Sigma = \left(\frac{R^2 \int |\nabla \times A|^2_{r=R} \sin\theta d\theta}{r_c^2 \int |\nabla \times A|^2_{r=r_c} \sin\theta d\theta} \right)^{\frac{1}{2}}. \quad (5.3)$$

Figure 5.2 shows a series of typical linear α^2 solutions with increasing diffusivity contrasts between the core and envelope. Linear mean-field dynamos of the α^2 type with a time-independent, spatially smooth scalar functional $\alpha(r)$ usually pro-

5.1 Early-Type Stars

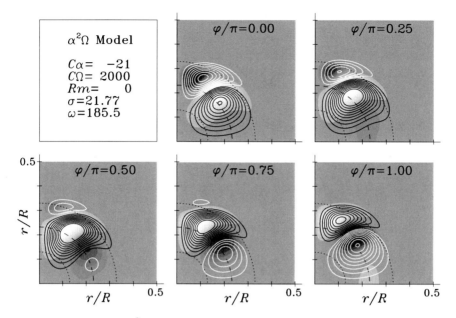

Fig. 5.3 A representative $\alpha^2\Omega$ solution. As this is an oscillatory solution, the eigenfunction is plotted at five equally spaced phase intervals ($\Delta\varphi = \pi/4$), covering half an oscillation cycle. The format in each panel is similar to Fig. 5.2. *White* (*black*) lines indicate fieldlines oriented in a clockwise (counterclockwise) direction. Note the wave-like propagation of the magnetic field from low to high latitudes. This symmetric solution has $C_\alpha = -21$, $C_\Omega = 2000$, $w/R = 0.1$, $\eta_e/\eta_c = 10^{-2}$, and is characterized by a growth rate $\sigma = 21.8\,\tau^{-1}$ and frequency $\omega = 186\,\tau^{-1}$. For $\eta_c = 10^9\,\mathrm{m^2\,s^{-1}}$, this corresponds to a dynamo period of about 7 yr, quite short compared to any other relevant timescales. Reproduced from Charbonneau and MacGregor (2001) by permission of the AAS.

duce *steady* magnetic fields, i.e., the solution eigenvalue is purely real ($\omega = 0$ in Eq. (3.46)). The solution plotted on Fig. 5.2a is dipole-like (i.e., antisymmetric), and is the fastest growing solution for our model with constant η, at the adopted value for C_α.[2] The poloidal and toroidal magnetic components have comparable strengths, which is again typical of α^2 mean-field models with scalar α-effect. Here the growth rate of the eigenmode is about 20 yr in dimensional units, leaving no doubt that ample time is available to amplify a weak seed magnetic field in the core of a massive star. Note also on Fig. 5.2 how the dynamo-generated magnetic field becomes trapped

---

[2] The α^2 form of the mean-field dynamo equations also admits growing solutions that are non-axisymmetric even though the α-effect profile exhibits axisymmetry with respect to the rotation axis. Growth rates for non-axisymmetric modes are often comparable to those of their axisymmetric counterparts. Motivated largely by the challenge posed by planetary magnetic fields, α^2 models can and have been constructed where non-axisymmetric modes are the fastest growing, and dominate in the moderately supercritical nonlinear regime. For complex enough spatial profiles of α, i.e., including multiple sign changes in each hemisphere, it is also possible to produce α^2 dynamo solutions undergoing cyclic polarity reversals.

within and in the immediate vicinity of the convective core for even moderately large values of magnetic diffusivity contrast between core and envelope.

In the presence of significant differential rotation, core dynamo action can produce polarity reversals and wave-like propagation of the magnetic field, much like in the $\alpha\Omega$ solar cycle models considered earlier. Figure 5.3 illustrates a half-cycle of a representative $\alpha^2\Omega$ solution, constructed by imposing a radial gradient of angular velocity across a thin shear layer coinciding with the core–envelope interface:

$$\Omega(r,\theta) = \Omega_c + \frac{\Omega_e - \Omega_c}{2}\left[1 + \mathrm{erf}\left(\frac{r-r_c}{w}\right)\right]. \tag{5.4}$$

The magnetic field distribution is shown at five distinct phases, at constant intervals of $\Delta\varphi = \pi/4$, in a format identical to that of Fig. 5.2 for each panel (note in particular that the eigenmodes are again plotted only in the inner half of the star). At a given phase, the solutions bear some resemblance to the α^2 solutions of Fig. 5.2c, in that the magnetic field is again trapped in the interior. As before, the toroidal field is concentrated near the core–envelope interface, and here, in fact, peaks slightly outside $r = r_c$ (dashed circular arc).

The availability of an additional energy source in the toroidal component of the dynamo equations leads to solutions where the toroidal field strength in general exceeds that of the poloidal field, scaling roughly as the ratio C_Ω/C_α in the limit $C_\Omega \gg C_\alpha$. For a given diffusivity ratio η_e/η_c, oscillatory $\alpha^2\Omega$ solutions have a smaller surface-to-core field strength ratio Σ than α^2 models, a direct consequence of the oscillatory nature of the field, which restricts the radial extent of the eigenfunction above the core–envelope interface to a distance comparable to the electromagnetic skin depth, which is very much smaller than the stellar radius for $\eta_e/\eta_c \ll 1$.

5.1.2 Numerical Simulations of Core Dynamo Action

It is interesting to compare and contrast core dynamo action, as modeled via mean-field electrodynamics, to what is produced by three-dimensional MHD simulations.[3] Such simulations do yield vigorous core dynamo action, with the magnetic energy approaching equipartition with the turbulent fluid motions. However, most of the magnetic energy is contained in small spatial scales, with the axisymmetric large-scale component accounting for only a few percent of the total magnetic energy. The simulations generate a highly time-variable differential rotation that contributes significantly to the induction of a toroidal component by shearing of the poloidal fields. This is most pronounced in the vicinity of the core–envelope boundary, where a persistent system of magnetic field bands approximately aligned in the azimuthal direction are produced.

---

[3] The content of this section is based primarily on the paper by Brun et al. (2005) given in the bibliography.

These simulations could be said to behave like $\alpha^2\Omega$ mean-field dynamo, but the analogy is only superficial because significant differences exist, most notably perhaps the absence of a well-defined, persistent mean-field-aligned turbulent electromotive force. Except in the innermost portion of the convective core, the mean kinetic helicity is negative in the Northern hemisphere, but in contrast the mean magnetic helicity does not show a well-defined, persistent hemispheric pattern, again a departure from mean-field expectations.

One important similarity with the mean-field models considered in Sect. 5.1.1 is the trapping of the magnetic field within or in the immediate vicinity of the convective core. This is shown on Fig. 5.4, which depicts two temporal and azimuthal averages at different epochs in a representative simulation. The weak axisymmetric toroidal field present in the inner portion of the radiative envelope is produced primarily by the shearing effect of the differential rotation, which is imprinted from the core to the lower envelope by the relatively high viscous forces characterizing this simulation.

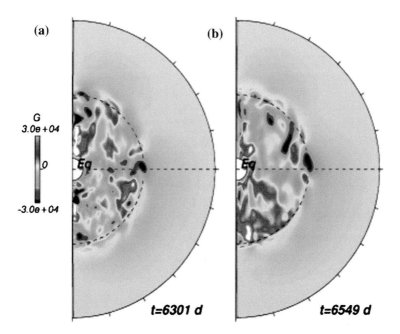

Fig. 5.4 Temporal+azimuthal average of the toroidal magnetic field in a 3D MHD numerical simulation of dynamo action in the core of a 2 $M_\odot$ early A-star. The magnetic field reaches a few tesla in strength, evolves rapidly, and is structured on a broad range of spatial scales, but remains confined to the convective core (*dashed circular arc*; the simulation domain only covers the inner 30% in radius of the star). Figure taken from Brun et al. (2005) (Fig. 18) by permission of the AAS.

5.1.3 Getting the Magnetic Field to the Surface

Whatever the mode of core dynamo action, a universal feature is the "trapping" of the magnetic field within the core and in the lower part of the radiative envelope, a direct consequence of the difficulty experienced by a magnetic field to diffusively penetrate a good electrical conductor. This is a long-recognized property of stellar core dynamos, and represents a rather formidable obstacle to be bypassed if the magnetic fields generated by dynamo action in the convective core are to become observable at the stellar surface. For O and B main-sequence stars, estimates for the diffusion time yield values largely in excess of the main-sequence lifetime. Introducing thermally-driven meridional circulation in the radiative envelope, expected to be a significant internal flow in rapidly rotating stars, does accelerate the transport of the deep field to the surface, but also impedes dynamo action.

Another possibility is that the dynamo-generated magnetic field manages to produce toroidal flux ropes that subsequently rise buoyantly to the surface. The analogy with the sun becomes even more compelling if a rotational shear layer does exist at the boundary between the inner convective core and outer radiative envelope. However, and unlike in the solar case, here the toroidal flux ropes are rising through a *stably* stratified environment, and so lose their buoyant force as they rise, because they cool faster than the surrounding stratification. Calculations performed in the thin flux tube approximation suggest that such toroidal flux ropes, assuming they do form, could rise perhaps halfway across the radiative envelope, but are unlikely to make it all the way to the surface through buoyancy alone. References listed in the bibliography should provide helpful entry points into the literature to those interested in further pursuing this aspect of massive star magnetism.

5.1.4 Alternative to Core Dynamo Action

Other dynamo-based explanations for the magnetic fields of early-type main-sequence stars certainly exist. One intriguing possibility that clearly requires serious modelling is that dynamo action in the outer layers of massive stars could take place in convection zones associated with a peak in iron opacities. Recent years have also witnessed renewed interest in the possibility that dynamo action could take place in the radiative envelope of intermediate- and high-mass main-sequence stars, through turbulence associated with one or more global instabilities of the magnetic field. This idea has attracted attention outside of the dynamo circles because the associated turbulent transport would also cause enhanced chemical mixing, known to be required to properly fit evolutionary tracks of massive stars, but whose origin remains mysterious. Introduction of simple parameterizations for the associated chemical and angular momentum mixing in models of evolving massive stars has shown that it is difficult to maintain sufficient differential rotation for the instability to operate, while keeping chemical mixing at the required level. The interested reader will find entry point in this vast literature in the bibliography at the end of this chapter.

This instability-driven dynamo mechanism probably cannot explain the strong magnetic fields observed in the very slowly rotating Ap/Bp stars (more on these presently), again because significant internal differential rotation is unlikely in the radiative envelope of these stars.

5.2 A-Type Stars

5.2.1 Observational Overview

Extant observations suggest a true dichotomy with regards to stellar magnetism in intermediate-mass stars: most A and B stars (around 95%) on or near the main-sequence have no measurable (as yet) magnetic field, but nearly all those that have, combine strong, large-scale magnetic fields, steady on decadal timescales at least, with slow rotation and pronounced photospheric abundance anomalies. As we will see later in this chapter, the presence of a strong, large-scale photospheric magnetic field (of whatever origin) favors angular momentum loss, and therefore slow rotation; and a strong magnetic field and slow rotation favor atmospheric stability, giving full leeway for chemical separation to operate and alter photospheric abundances.

In the most slowly rotating, strongly magnetized Ap stars, the mean surface magnetic field strength ("mean" in the sense of being averaged over the stellar surface) can be detected by the Zeeman splitting of spectral lines, as in sunspots. Figure 5.5 below shows a striking example of such splitting. In more rapidly rotating stars, magnetic Doppler imaging becomes a possibility; this relies on the varying shapes of spectral lines formed as magnetic structures cross the visible part of the stellar disk. "Imaging" remains indirect, in the sense that the stellar surface is of course not resolved spatially, but the availability of many spectral lines, with some appropriate regularization scheme, allows this inverse problem to be solved. Figure 5.6 shows a particularly well-studied exemplar, namely the chemically peculiar star 49Cam. The field strength is high, the magnetic topology quite complex, with the idea of a strongly inclined dipole, historically the common interpretation for Ap stars magnetic fields, being a rather rough approximation here.

It is an intriguing fact that the few chemically-normal, (relatively) rapidly rotating early-type stars on which magnetic fields have been detected all sit in the early-B range of spectral types and belong to the βCep sub-class (and include the prototype star βCep itself). However, indirect evidence for photospheric magnetism in O and B stars has been accumulating steadily, be it as emission of hard radiation above and beyond what shock dissipation can provide, channelling of stellar winds, and spectral variability. Ongoing spectropolarimetric campaigns targeting massive stars are now providing more and more data for theoreticians/modellers to chew on in upcoming years.

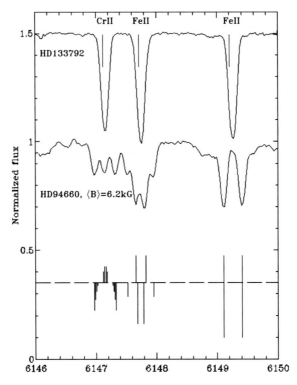

Fig. 5.5 Zeeman splitting of magnetically sensitive absorption lines in the spectrum of the Ap star HD94660. The inferred mean field strength for this star is 0.62 T. The *top* trace is that of a typical unmagnetized star of similar spectral type. The horizontal axis is the wavelength, measured in Å. The *bottom* trace shows the multiplet structure of the three spectral lines. Figure reproduced from the Mathys et al. (1997) paper cited in the bibliography, with a few labels added. With permission © ESO.

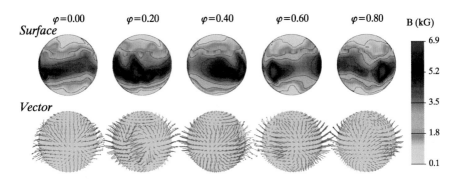

Fig. 5.6 The surface magnetic field on the Ap star 49 Cam, as reconstructed for various rotational phases (φ) by magnetic Doppler imaging. The *top row* shows the net field strength, and the *bottom row* the orientation of the surface magnetic field vector. Plot courtesy of J. Silvester and G. Wade, RMC/Kingston, taken from Silvester et al. (2009) by permission of Cambridge University Press.

5.2.2 The Fossil Field Hypothesis

Stars with spectral types ranging from late-B to early-F stand out as the least likely to support dynamo action, because they lack a convective region of substantial size. This squares well with various lines of observations; in particular, main-sequence A-stars are amongst the most "magnetically quiet" stars in the HR diagram. As we just discussed, a subset of late-B and A stars, namely the slowly-rotating, chemically peculiar Ap/Bp stars, do show strong magnetic fields, but show no sign of anything even mildly analogous to solar activity. The single pattern of temporal evolution noted is a decrease by factors of 2–3 in the overall strength of the surface field, most prominent in the early stages of main-sequence evolution. This seems compatible with the idea of diffusive decay of residual higher-degree eigenmodes, and slow decreases associated with flux conservation as the stars slowly expand in the course of their main-sequence evolution (cf. Sect. 2.1).

5.2.3 Dynamical Stability of Large-Scale Magnetic Fields

The study of the purely resistive decay of large-scale magnetic fields in stellar interiors carried out in Sect. 2.1 precluded, by it very design, the development of flows propelled by potential hydromagnetic instabilities. Investigations into the latter have shown that under typical stellar interior conditions, large-scale magnetic fields embedded in stably-stratified radiative interiors are indeed susceptible to the development of instabilities with growth rates much smaller than any relevant evolutionary timescales. Even simple field configurations, such as low-order multipole purely toroidal or purely poloidal fields are found to be unstable, with rotation possibly providing a stabilizing influence at high rotation rates (see references in the bibliography at the end of this chapter).

Although these (linear) stability analyses rely on a number of strong simplifying assumptions, they lead to a picture whereby the most likely stable global configurations are magnetic fields comprised of a mixture of large-scale poloidal and toroidal components of comparable strengths. Remarkably, this has recently been confirmed by full MHD simulations. Such configurations would establish themselves on very short timescales, after which they would undergo resistive decay on the magnetic diffusion timescale. Short-lived unstable phases early in their evolution notwithstanding, this overall picture remains generally consistent with the fossil field Ansatz for Ap/Bp stars.

5.2.4 The Transition to Solar-Like Dynamo Activity

On the main-sequence, as one moves down from late-A to late-F spectral types, solar-type surface convection sets in, with the convection zone rapidly gaining in depth below spectral type F5. How and when solar-type dynamo action sets in is a relatively unexplored question that clearly deserves further attention, from both the observational and modelling standpoints.

5.3 Solar-Type Stars

5.3.1 Observational Overview

The photosphere of solar-type stars other than the sun cannot be spatially resolved, and so direct observation of starspots is not possible, although rotational modulations of the luminosity associated with starspot darkening most certainly is. Direct measurements of magnetic polarisation of starlight is difficult as well, unless the field has a strong large-scale component, otherwise the polarisation associated with regions of opposite polarities—e.g., starspot pairs—cancel out when integrated over the solar disk. Most evidence for the presence of magnetic fields on such stars is thus indirect, yet extremely compelling, as it covers a wide range of phenomena visible on the sun, such as spectral line variability, rotational modulation of luminosity due to the passage of large starspots, flares, radio bursts, and variability in magnetically-sensitive spectral lines on a wide range of timescales. Such indirect observational evidence for magnetic fields has been found on *every* late-type main-sequence star observed with sufficient sensitivity. The sun is, indeed, a typical solar-type star!

One stellar observable that is particularly noteworthy is the emission in the cores of the H and K lines of CaII (396.8 and 393.4 nm, respectively). On the sun, this emission is known to be associated with non-radiative heating of the upper atmosphere, and is known to scale well with the local photospheric magnetic flux. Starting back in 1968 at Mt. Wilson Observatory, Olin C. Wilson (1909–1994) began measuring the CaII H+K flux in a sample of solar-type stars, a laborious task that was later picked up by a brave group of undeterrable associates and followers, whose collective labor has produced a 40+ year long archive of CaII emission time series for no less than 111 stars in the spectral type range F2–M2, on or near the main-sequence.

Figure 5.7 shows a few sample time series of the so-called Calcium index S, measuring the ratio of core emission intensity in the H and K lines to that of the neighbouring continuum. Some stars show solar-like cycles, others have irregular CaII emission, some show long term trends and others can only be dubbed "flatliners". Note that the mere presence of detectable CaH+K emission indicates magnetic activity; the absence of detectable temporal variations in flatliner stars simply means that these stars lack a solar-like large-scale magnetic field undergoing cyclic variation.

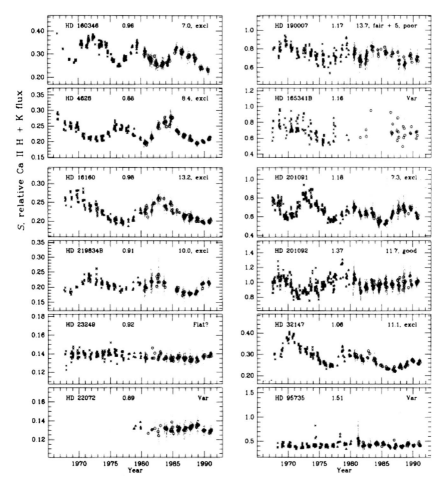

Fig. 5.7 Calcium emission index in a small subsample of the Mt. Wilson dataset, showing the variety of CaII emission patterns: cycles, non-cyclic irregular emission, long term trend, and constant emission. On such plots, the sun would have a mean emission level $\langle S_\odot \rangle = 0.179$, with a min/max range of about 0.04. Figure cropped from a much larger figure in Baliunas et al. (1995) (Fig. 1g) by permission of the AAS.

5.3.2 Empirical Stellar Activity Relationships

From the point of view of dynamo theory and modelling, the following two empirical facts are particularly noteworthy:

1. Magnetic activity, as measured e.g. by the level of CaH+K emission, generally increases with increasing rotation rate (decreasing $P_{\rm rot}$);
2. Stellar cycle periods increase with increasing rotation periods $P_{\rm rot}$.

Among stars showing cycles in CaH+K, at a given spectral type the relationship between cycle period and rotation rate is well represented by a power-law of the form $P_{\text{cyc}} \propto P_{\text{rot}}^n$, with n varying between 0.75 and 1.75 depending on spectral type. Interestingly, all these data can be described reasonably well by a single power-law fit to the ratio of the rotation period to the convective turnover time, known as the Rossby Number (Ro; the inverse Rossby number is often referred to as the Coriolis number):

$$P_{\text{cyc}} \propto \left(\frac{P_{\text{rot}}}{\tau_c}\right)^n, \quad n = 1.25. \tag{5.5}$$

Recall from the discussion in Sect. 3.2.2 that this ratio is supposed to measure the efficiency of the Coriolis force in breaking the mirror-symmetry of convective turbulence, and thus producing a non-zero α-effect. Recall also that the larger the dynamo number, the more magnetic energy mean-field models can produce (viz. Fig. 3.8). So, in a rough qualitative sense, observations seem to fit our (naive) theoretical expectations.

In reality, there are of course significant complications to this highly simplified picture. For example, coronal (X-Ray) and chromospheric (Ca H+K) emission is observed to saturate as Ro falls below about 0.1, and even decreases a bit beyond Ro $\sim 10^{-2}$, although it is not clear whether this reflects a saturation of the emission mechanism, of the surface filling factor of magnetized structured, or of magnetic field generation by the large-scale dynamo. The P_{rot} versus Ro relationship becomes a lot tighter if stars for which reliable cycle periods are known are first divided into "active" and "inactive" subgroups on the basis of their overall level of Ca emission.

Solar-like cyclic activity is by no means the rule among solar-type stars, with only about 60% of stars showing well-defined cycles, 25% showing aperiodic variations, and the remaining 15% being "flatliners" with low level, constant chromospheric emission. Indeed, the sun has "twins", i.e., main-sequence stars of closely similar surface temperature, gravity and rotation rate, which do not show any variability in chromospheric emission. An intriguing possibility is that these stars just happened to have been caught in a Maunder Minimum-like phase of suppressed cyclic activity.

5.3.3 Solar and Stellar Spin-Down

Stellar observations indicate that there is evidently a lot more to dynamo action than just rotation, nonetheless the latter is clearly a key factor. For this reason, in any attempt to secure a coherent picture of dynamo action in solar-type stars, an important global feedback mechanism of dynamo action must first be considered: angular momentum loss, and its effect on stellar rotation rates.

Although the existence of systematic differences between the average rotation rates of early- and late-type stars was known already for nearly a hundred years,

5.3 Solar-Type Stars

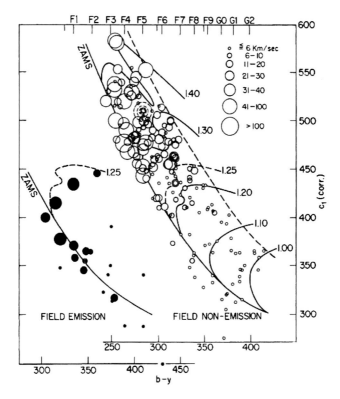

Fig. 5.8 Distribution of projected rotational velocities ($v \sin i$) for main-sequence stars, plotted in an observational HR diagram. Luminosity increases vertically upwards, and effective temperature horizontally leftward. Astronomical spectral types are listed along the upper axis. *Solid lines* are stellar evolutionary tracks, labeled according to mass in solar units. These tracks, particularly for $M/M_\odot \gtrsim 1.2$, are now somewhat obsolete. Diagram reproduced from Kraft (1967) (Fig. 1) by permission of the AAS.

observational evidence for main-sequence spin-down of solar-type stars was established much later. Figure 5.8 below is a reproduction of a diagram put together by Robert Kraft in 1967, showing the distribution in a HR diagram of projected equatorial rotational velocities ($v \sin i$) measured in a sample of field stars. As one runs down the main sequence, there occurs a sharp drop in $v \sin i$ starting around spectral type F5. Slow rotation is the rule on the cool side of this so-called rotational dividing line, while on the hot side rapid rotation is common. Kraft went on to show that under the assumption of solid-body rotation, in the interval $1.5 \lesssim M/M_\odot \lesssim 20$ observed rotation rates are consistent with a power-law dependence between stellar angular momentum (J) and mass (M) of the form $J \propto M^{1.57}$, which abruptly breaks down below F5.

The decrease in the moment of inertia of stars associated with their contraction towards the main-sequence can easily account for ZAMS equatorial rotational veloc-

Fig. 5.9 Main-sequence temporal evolution of rotation rates, Calcium emission and Lithium abundance in solar-type stars. Diagram reproduced from Skumanich (1972) (Fig. 1) by permission of the AAS.

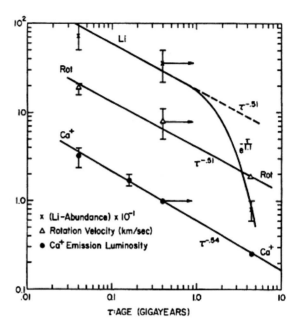

ities of a few hundreds of kilometers per second. As was already understood then, the anomaly in Kraft's diagram lay in fact with the slowly-rotating low-mass stars. A most spectacular illustration that this is due to main-sequence *spin-down* and is associated with magnetism was provided in a short, now classical 1972 paper by Andrew Skumanich. Figure 5.9, reproduced from this paper, illustrates the simultaneous and gradual decrease of both the average rotation rates and magnetic activity for late-type stars—as measured by emission in the core of the Ca H and K lines—in a few open clusters of known ages. Later observations focusing on young clusters such as αPersei and the Pleiades have revealed that main-sequence spin-down for late-type stars is very swift, with the bulk of it completed in the first few 100 Myr after arrival on the ZAMS.

The key in explaining main-sequence spin-down is the realisation that stars with hot coronae lose mass through thermally-driven winds, and that the presence of a coronal magnetic field—ultimately of dynamo origin—turns out to greatly enhance the loss of angular momentum in the wind. We first examine this issue, using a geometrically simple but dynamically self-consistent MHD wind solution known as the Weber–Davis (WD) model.

Working under ideal MHD, we consider a steady ($\partial/\partial t = 0$) spherically symmetric ($\partial/\partial \theta = \partial/\partial \phi = 0$) compressible outflow from a star rotating at angular velocity Ω and characterized by a known surface radial component of the magnetic field B_{r0}. The coronal base temperature $T(r_0) \equiv T_0$ is considered given, and the energy equation is summarily dealt with by assuming a polytropic relationship between pressure and density. Outflow solutions are sought only in the equatorial

5.3 Solar-Type Stars

plane, where we also set $B_\theta = 0$. This may smell of monopolar magnetic fields, but this is actually a fair representation of the interplanetary magnetic field measured in the ecliptic plane during solar minimum conditions. Under these assumptions, mass conservation and the $\nabla \cdot \boldsymbol{B} = 0$ constraint yield two conservation statements for the mass and magnetic flux across a spherical surface:

$$\frac{1}{r^2}\frac{\partial}{\partial r}(r^2 \varrho u_r) = 0 \quad \rightarrow \quad r^2 \varrho u_r = C_1, \tag{5.6}$$

$$\frac{1}{r^2}\frac{\partial}{\partial r}(r^2 B_r) = 0 \quad \rightarrow \quad r^2 B_r = C_2, \tag{5.7}$$

where C_1 and C_2 are integration constants, corresponding respectively to the mass and magnetic flux carried by the wind. The ϕ-component of the induction equation is also readily integrated to yield:

$$\frac{1}{r}\frac{\partial}{\partial r}(r u_r B_\phi - r u_\phi B_r) = 0 \quad \rightarrow \quad r(u_r B_\phi - u_\phi B_r) = C_3. \tag{5.8}$$

To evaluate the integration constant C_3 we transform to a reference frame co-rotating with the star, i.e., $u_\phi \rightarrow u'_\phi + \Omega r$, where the prime indicates evaluation in the co-rotating frame. Note that this (non-relativistic) transformation leaves the radial components of $\boldsymbol{u}$ and $\boldsymbol{B}$ unaffected. In that frame $\boldsymbol{B}$ is stationary, and since we are working under the flux-freezing approximation $\boldsymbol{u}$ and $\boldsymbol{B}$ must be parallel: $u'_r/u'_\phi = B'_r/B'_\phi$. Since $B_r = B'_r$ and $u_r = u'_r$, and remembering that the magnetic field is assumed purely radial at the base of the wind (so that $B_\phi = B'_\phi = 0$ at $r = r_0$), Eq. (5.8) yields $C_3 = -\Omega r^2 B_r$, so that

$$B_\phi = \frac{B_r}{u_r}(u_\phi - \Omega r). \tag{5.9}$$

Now, under the geometry and flow symmetry considered here, the ϕ-components of the momentum equation can be brought to the form:

$$\frac{\partial}{\partial r}(r u_\phi) = \frac{B_r}{\mu_0 \varrho u_r}\frac{\partial}{\partial r}(r B_\phi); \tag{5.10}$$

but in view of Eqs. (5.6) and (5.7), we have $B_r/\mu_0 \varrho u_r = C_2/\mu_0 C_1$, i.e., a constant! Consequently, Eq. (5.10) integrates immediately to

$$r u_\phi - \frac{r B_\phi B_r}{\mu_0 \varrho u_r} = L, \tag{5.11}$$

where L is yet another integration constant. It has a well-defined physical meaning, as it corresponds to the total specific angular momentum carried away by the wind, which is made up of two contributions: the specific angular momentum of the expanding fluid (first term on LHS), and the torque density associated with magnetic tension.

Using Eq. (5.9) to substitute for B_ϕ, and expressing the magnetic field components in terms of the corresponding Alfvén velocity components (Sect. 1.8):

$$a_r = \frac{B_r}{\sqrt{\mu_0 \varrho}}, \quad a_\phi = \frac{B_\phi}{\sqrt{\mu_0 \varrho}}, \tag{5.12}$$

produces, after some straightforward algebra:

$$u_\phi = \Omega r \frac{(u_r^2 L/\Omega r^2) - a_r^2}{u_r^2 - a_r^2}. \tag{5.13}$$

The denominator of this expression vanishes if the radial flow velocity ever becomes equal to the radial Alfvén speed, unless the numerator also happens to vanish. Regularity of the solution through this critical point then requires that we set

$$\boxed{L = \Omega r_A^2}, \tag{5.14}$$

where r_A is the *Alfvén radius*, defining the spherical shell where $u_r = a_r$. Now, remember that L is the total angular momentum carried away by the wind, *including* the torque density provided by magnetic tension. Equation (5.14) states that this is equal to the angular momentum that would be carried away by an unmagnetized wind flowing strictly radially, and co-rotating with the solar/stellar surface out to radius r_A.

Equation (5.14) holds only in the equatorial plane, where the WD solution is computed. The WD model can be "stretched" to the whole sphere by assuming that a whole spherical shell is co-rotating out to r_A; this means replacing Eq. (5.7) by:

$$L_{\text{sph}} = \frac{2}{3} \Omega r_A^2, \tag{5.15}$$

where the factor 2/3 arises from the moment of inertia integral. The angular momentum loss rate then follows directly from multiplication by the mass loss rate:

$$\frac{dJ}{dt} = \dot{M} \times L_{\text{sph}} = -4\pi \varrho_A r_A^2 u_{rA} \left(\frac{2}{3} \Omega r_A^2\right). \tag{5.16}$$

At the Alfvén radius we have $u_{rA} = a_{rA}$, with $B_{rA}^2 = 4\pi \varrho_A a_{rA}^2$. Moreover, conservation of magnetic flux implies $r_0^2 B_{r0} = r_A^2 B_{rA}$. Putting all this into Eq. (5.16) leads to

$$\frac{dJ}{dt} = -\frac{2}{3} B_{r0}^2 r_0^4 \Omega a_{rA}^{-1}. \tag{5.17}$$

5.3 Solar-Type Stars

Knowing the stellar moment of inertia I and assuming rigid rotation throughout the interior, the spin-down timescale is readily calculated:

$$\tau_{sp} = I\Omega \left(\frac{dJ}{dt}\right)^{-1}. \tag{5.18}$$

Now, for rotating magnetized winds that are mostly thermally driven, a_{rA} is of the order of the sound speed ($c_s = \sqrt{kT/\mu m_p} \sim 10^5$ m s^{-1} for a coronal temperature of $\sim 10^6$ K) to within a factor of two or so. If the coronal temperature is held fixed, this means that the angular momentum loss rate is only a function of the rotation rate and surface magnetic field strength.

We encountered in earlier chapters various lines of argument indicating that the dynamo-generated magnetic field strength should increase with increasing rotation rate, an expectation also buttressed by observations of chromospheric activity in solar-type stars of varying rotation rates. If one assumes $B_{r0} \propto \Omega$, and for a fixed moment of inertia on the main-sequence (the latter a very good approximation, for a change...), Eq. (5.17) then lead to

$$\frac{d\Omega}{dt} \propto -\Omega^3. \tag{5.19}$$

This already indicates that faster rotating stars spin down a lot faster than their more slowly rotating cousins, which provides a natural explanation for the convergence of rotation rates observed at a given spectral type when looking at stellar rotation in progressively older clusters. Now, Eq. (5.19) readily integrates to.

$$\frac{1}{\Omega^2(t)} - \frac{1}{\Omega^2(t_0)} \propto t - t_0, \tag{5.20}$$

where t_0 is the time of arrival on the ZAMS (or shortly thereabouts). In the asymptotic limit $t \gg t_0$, $\Omega \ll \Omega(t_0)$, this becomes

$$\Omega(t) \propto t^{-1/2}, \tag{5.21}$$

which, how about that, is precisely the power-law relationship inferred observationally by Skumanich (cf. Fig. 5.9). For the sun, with $I \simeq 10^{47}$ kg m^2, $\Omega = 2.6 \times 10^{-6}$ rad s^{-1} and $B_0 \sim 2 \times 10^{-4}$ T one finds a leisurely spin-down timescale of about 5 Gyr; but in a "young sun" with a rotation period of 2 day and $B_0 = 3 \times 10^{-3}$ T, this drops to a few 10^7 yr, indicating that rapidly rotating young solar-type stars spin down swiftly after arriving on the ZAMS.

5.3.4 Modelling Dynamo Action in Solar-Type Stars

The above discussion indicates that one could expect dynamo action to be far more vigorous in young, rapidly rotating solar-type stars, and the good fit of our spin-down model prediction with Skumanich's $t^{-1/2}$ relationship even suggests a linear increase of magnetic field strength with rotation rate, at least up to ~ 10 times the present solar rotation if coronal X-Ray emission can be taken as proxy of dynamo efficiency. Can such trends be convincingly recovered from the various solar dynamo models introduced in Chap. 3? In practice, we are facing a number of difficulties in carrying out such an "extrapolation" to stars other than the sun, with convection zones of greater or lesser depths, differing luminosity, and a range of rotation rates. At the very least we need to be able to specify:

1. How the form and magnitude of differential rotation and meridional circulation change with rotation rate and luminosity, the latter determining the magnitude of convective velocities, and thus the magnitude of the turbulent Reynolds stresses powering the large-scale flows important for dynamo action;
2. How the α-effect and turbulent diffusivities vary in stars with different rotation rates and convection zone properties;
3. How the process of sunspot formation (essential in Babcock–Leighton models) varies with varying convection zone depth, rotation, etc.

It is quite sobering to reflect upon the fact that we currently do not have theories or models allowing us to provide firm, quantitative and robust answers to *any* of these questions. Moreover, the preceding two chapters should have made it clear that even in the sun, we don't really know for sure what is the mechanism responsible for the regeneration of the poloidal magnetic component. How then can we hope to go about modelling stellar dynamos with anything resembling confidence? At this point in time I would argue that we cannot. However, the problem can be turned around, in that stellar observations can perhaps be used to distinguish between different classes of dynamo models.

The possibility hinges on features like the distinct dependency of the cycle period on model parameters in various models. For the simple α-quenched kinematic mean-field dynamo solutions discussed in Sect. 3.2.9, the (dimensionless) cycle period is, to a first approximation, independent on the dynamo numbers (see Fig. 3.6b), so that the physical period scales primarily as

$$P_{\text{cyc}} \propto \eta_0^{-1} \quad [\alpha\text{-quenched } \alpha\Omega \text{ model}], \tag{5.22}$$

where η_0 is the assumed turbulent diffusivity. Mean-field models including more complex form of nonlinearities produce more complex parametric dependencies, but a strong dependency on η always emerges. This quantity, in turn, is expected to increase with increasing convective velocities (cf. Eq. (3.22)), and therefore with increasing luminosity. On the other hand, in Babcock–Leighton dynamo models operating in the advection-dominated regime, the cycle period is found to be controlled primarily by the turnover time of the meridional flow cell, with a much weaker

5.3 Solar-Type Stars

dependency on the assumed value for the turbulent diffusivity. For the specific "solar" model described in Sect. 3.3, the cycle period is found to vary as:

$$P_{\text{cyc}} \propto u_0^{-0.89} s_0^{-0.13} \eta_0^{-0.22} \quad \text{[Babcock–Leighton]}, \tag{5.23}$$

where u_0 is the surface meridional flow speed (see Fig. 3.10, top left), and s_0 is the parameter measuring the magnitude of the Babcock–Leighton source term in Eq. (3.64).[4] Unfortunately, using this relationship in conjunction with observed stellar cycle data requires one to specify how the meridional flow speed varies with rotation, which currently remains highly uncertain on the theoretical and simulation fronts. But this is a very promising avenue.

All of these model-based cycle period formulae have been obtained in kinematic dynamo models, and how well they hold in the dynamical nonlinear regime is entirely unknown. With regular cycles now reproduced in global 3D MHD simulations of solar convection (cf. Sect. 3.5), these dynamical effects may become ripe for investigation in the near future.

5.4 Fully Convective Stars

We now move to the bottom end of the main-sequence, where stars become fully convective around spectral type M5. Observationally, no obvious discontinuity is observed in X-Ray or Ca H+K emission as one moves into spectral types M, and indeed some of the more active (single) flare stars are fast rotators of very late spectral type. One can but conclude that a dynamo of some form is still operating.

One might have expected dynamo action in fully convective stars—either in late-M main-sequence or pre-main sequence TTauri stars—to be a mere variant on core dynamo action in massive stars, but in fact a number of significant differences come into play, related to the physical conditions at the boundary of the convecting sphere. Full-sphere MHD simulations[5] of a "M-star in a box" by Dobler et al. (2006) are particularly interesting in this respect. They indicate that vigorous dynamo action does occur, with the magnetic energy at $\sim 20\%$ of equipartition with the turbulent fluid motions at low to moderate rotation, and reaching equipartition at high rotation rates. The simulations are characterized by a very well-defined, persistent spatial pattern of mean kinetic helicity, again negative in the N-hemisphere (see Fig. 5.10a). This leads to the production of a well-defined large-scale magnetic component, with energy content going from some 20% of the total magnetic energy at low rotation, up to $\sim 50\%$ at high rotation rates. The large-scale field has poloidal and toroidal

---

[4] Note however that the above relation was calibrated in a relatively narrow range of parameters: $2 \leq u_0 \leq 30\,\text{m s}^{-1}$, $0.03 \leq s_0 \leq 1\,\text{m s}^{-1}$, $2 \times 10^6 \leq \eta_0 \leq 5 \times 10^7\,\text{m}^2\,\text{s}^{-1}$, and is only expected to hold in the so-called advection-dominated regime; see the paper by Dikpati & Charbonneau (1999) cited in the bibliography of Chap. 3.

[5] The content of this section is based primarily on the paper by Dobler et al. (2006).

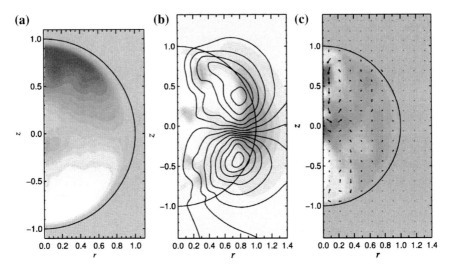

Fig. 5.10 Temporal+azimuthal average of **a** kinetic helicity, **b** toroidal and poloidal magnetic field, and **c** large-scale flows in a complete MHD simulation (including overall structure) of a fully convective star including central heat source and surface cooling. The *gray scale* codes kinetic helicity in part **a**, the toroidal magnetic component in **b**, and the zonal flow component in **c**, *gray to black* (*white*) coding negative (positive) values. Figure adapted from Dobler et al. (2006) (Figs. 5 and 10) by permission of the AAS.

components of comparable strengths, typical of mean-field α^2 dynamos, and is often dominated by a well-defined quadrupolar component (see Fig. 5.10b). Because the convecting sphere cannot exchange angular momentum across its outer boundary (here the stellar surface), differential rotation is much weaker than in the massive star core dynamo simulations reviewed in Sect. 5.1.2), and is concentrated in the vicinity of the rotation axis, as shown on Fig. 5.10c).

Much like in the core dynamo simulations briefly described in Sect. 5.1.2, these simulations only reach moderate values of the viscous and magnetic Reynolds numbers, many orders of magnitude below what one would expect under stellar interior conditions. Nonetheless the important bottom line, once again, is that production of magnetic field through dynamo action here also appears inescapable.

5.5 Pre- and Post-Main-Sequence Stars

As with main-sequence late-type stars, abundant evidence for magnetic fields in pre- and post-main sequence stars of spectral types later than F has now been accumulating, mostly again in the form of stellar analogs to well-observed solar phenomena: X-Ray and EUV emission, flaring, spectral variability, rotational modulation by starspots, and so on. More recently, magnetic Doppler imaging has been used to reconstruct the surface magnetic field of some pre-main-sequence stars in the TTauri evolutionary phase. Whether TTauri or giants, all these stars have low surface temperature and thick convection zones, so observations of magnetic activity indicators similar to what is observed in late-type main-sequence stars point once again to the importance of convection zones of significant radial extent below the photosphere. Indeed, there seems to exist a rather clear-cut, slightly inclined dividing line bisecting the upper part of the HR diagram (main-sequence and up in luminosity), on the right side (low T_{eff}) of which evidence of magnetic activity is ubiquitous. Things get messy again with very cool supergiants, with signs of magnetic activity disappearing across various not quite coincident dividing lines, depending on the indicator chosen (X-Ray emission, non-thermal emission lines, etc).

With classical TTauri stars, additional complications also come from the presence of an accretion disk, itself most likely magnetized, perhaps the site of magnetic field generation by dynamo action, and perhaps even magnetically coupled to its central star. Such a coupling has been invoked to explain the (relatively) low rotation rates of TTauri stars, which after all are contracting and accreting large amounts of mass—and angular momentum—from their disk, and should therefore spin up far more than is observed. Indeed, without angular momentum loss mediated by magnetic fields in the early stages of star formation, it is quite likely that stars could simply not eliminate enough angular momentum to form at all!

In hot post-main sequence stars, the observational situation is not well documented or understood. It is a remarkable fact that magnetic fields have been detected in *all* sdO and sdB hot subdwarfs for which a serious attempt has been made. The evolutionary status of these objects is not well-understood, but they most likely represent what used to be the inner core of giants prior to the episode of strong mass loss that accompanies the transition to the horizontal branch. Detection of ~ 0.1 T-strength magnetic fields in such stars is strong evidence for the existence of magnetic fields in the deep interior of their main-sequence progenitors.

5.6 Compact Objects

Magnetic fields in isolated white dwarfs have been detected through circular polarisation measurements in the wings of strong spectral lines, usually Balmer lines in the so-called "DA" white dwarfs showing Hydrogen lines in their photospheres. Actual Zeeman splitting is only detected in the most strongly magnetized objects ($\gtrsim$ a few 10^2 T). Inferred field strengths range from a few T up to a whopping 10^4 T, with a few objects possibly approaching 10^5 T, and the overall occurrence of detectable magnetism standing at 10–15%. However, these techniques are only sensitive to large-scale magnetic fields, still producing a net polarisation signal when integrated over the stellar disk, and so the true incidence of magnetism in white dwarfs may actually be significantly higher. At around 22,000 K on the white dwarf cooling track, there appears a subclass of Carbon-rich objects, the so-called "hot DQ" white dwarfs, in which the Carbon enrichment is believed to be due to the onset of envelope convection and associated mixing. Interestingly, the incidence of magnetism in this subclass of white dwarfs is some four times larger than across the overall white dwarf population. Could this sudden increase in the incidence of magnetism be physically associated with the onset of envelope convection through a dynamo mechanism? This interesting question remains to be explored.

Inferred magnetic field strengths in neutron stars range from 10^4 to 10^{11} T. Neutron stars magnetic fields are of course most readily detected via the pulsar phenomenon, arising from the misalignment of the magnetic axis with respect to the rotation axis of the (very rapidly rotating) neutron star. It is quite striking that the highest strengths of large-scale magnetic fields in main-sequence stars (a few T in Ap stars), in white dwarfs ($\sim 10^5$ T) and in the most strongly magnetized neutron stars ($\sim 10^{11}$ T) all amount to similar surface magnetic fluxes, lending support to the idea that these high field strengths can be understood from simple flux-freezing arguments (Sect. 1.10). There is also observational evidence that actual magnetic field evolution is taking place as pulsars age, but this remains very slippery territory, both from the modelling and observational points of view.

Observationally, very little is known about black holes except that there seems to be one at the center of our galaxy, so you won't be surprised to hear that even less is known about black hole magnetic fields. One should perhaps just point out that solutions to the field equations of general relativity for rotating, electrically charged black holes do exist, which is a good start towards magnetic fields production. Evidence to date is limited to energetic phenomena interpreted in terms of magnetic channelling of material onto the black hole. But beyond that, at the present time, there is mostly pure fervor.

5.7 Galaxies and Beyond

Magnetic fields in the diffuse, low-density interstellar gas is most readily detected through synchrotron radiation emitted by relativistic charged particles spiralling along magnetic fieldlines. This technique is successful not only within the Milky Way, but also for other galaxies. Other means of detection, for the time being limited to the Milky Way, include the polarisation of optical starlight by elongated (i.e., non-spherical) dust grains aligning themselves perpendicularly to magnetic fieldlines; these aligned dust grains also sometimes emit detectable polarized infrared radiation. Finally, for relatively strong fields, Zeeman splitting of spectral lines in the radio domain has also been measured. As with stars, magnetic fields seem to be ubiquitous features in pretty much all galaxies. Indirect evidence for the existence of extragalactic magnetic fields also exists, with an upper limit of $\sim 10^{-3}$ nT on the mean field strength over length scales of order 100 MPc and larger. These fields could be primordial in origin, or could have been ejected into intergalactic space by galactic winds.

The galactic magnetic field in the solar neighbourhood has a strength of about 0.6 nT, up to a few nT near galactic center. This is indeed typical of spiral galaxies, which show field strengths in the range 0.5–1.5 nT, up to some 3 nT in high density regions of spiral arms. The strongest large-scale galactic magnetic fields so far measured have strength reaching 1 nT, and have been found in starburst galaxies. While this may seem quite low values, such field strengths have important consequences for star formation, the distribution of cosmic rays, and equilibrating the interstellar medium against gravity.

Given that most stars appear to be magnetized to some degrees, and that many stars tend lose mass (some by blowing up!), it is perhaps not surprising to detect magnetic field in the galactic interstellar medium. What is surprising is that this magnetic field tends to be organized on large spatial scales, commensurate in fact with galactic dimensions. Such large-scale, spatially well-organized magnetic fields are most likely produced by a dynamo mechanism, not at all dissimilar to that responsible for the presence of magnetic fields in many stars, including the sun. Simple kinematic models, much like those considered in Sect. 3.2, have been built, in which differential rotation in the disk and interstellar turbulence driven by supernovae explosion and/or magnetic buoyancy instabilities could jointly act as a $\alpha\Omega$ dynamo on galactic scales. Additional, indirect evidence for well-organized large-scale magnetic fields in galaxies include the collimation of jets, and energetic phenomena often encountered in quasars and AGN; at the present time, the most convincing physical models for such phenomena all involve magnetic fields at some level.

These galactic magnetic fields can also provide the seed required to start up the dynamo processes in the sun and stars, as per the linearity of the MHD induction equation, which brings us back almost to the beginning of our grand tour of solar and stellar dynamos. The next lap is yours to take. Have fun with it!

Bibliography

> The following conference proceedings gives an excellent sampling of the current state-of-the-art in the observation of stellar magnetic fields:

Neiner, C., & Zahn, J.-P., eds.: 2009, *Stellar Magnetism*, EAS Publications Series, 39

> The first three chapters therein, authored by John Landstreet (pps. 1–53), provide an outstanding overview of the subject, in just a little over 50 pages.

> The linear α^2 and $\alpha^2\Omega$ models for core dynamo action in massive stars presented in Sect 5.1 are taken pretty directly from

Charbonneau, P., & MacGregor, K.B.: 2001, *Magnetic fields in massive stars. I. Dynamo models*, Astrophys. J., **559**, 1094–1107

> This paper also adresses the effects of thermally-driven meridional circulation in the radiative envelope, including its deleterious effect on core dynamo action. The difficulty in bringing the magnetic fields produced by core dynamo action to the surface of a star with a thick radiative envelope was cogently demonstrated a while ago already by

Levy, E.H., & Rose, W.K.: 1974, *Production of magnetic fields in the interiors of stars and several effects on stellar evolution*, Astrophys. J., **193**, 419–427

Schüssler, M., & Pähler, A.: 1978, *Diffusion of a strong internal magnetic field through the radiative envelope of a 2.25-solar-mass star*, Astron. & Astrophys., **68**, 57–62

The discussion of core dynamo action in massive stars (Sect. 5.1.2) is based on the following paper:

Brun, A.S., Browning, M.K., & Toomre, J.: 2005, *Simulations of core convection in rotating A-type stars: magnetic dynamo action*, Astrophys. J., **629**, 461–481

from which Figure 5.4 was taken. For calculations of buoyantly rising thin flux tubes in the radiative envelope of massive stars, see

MacGregor, K.B., & Cassinelli, J.P.: 2003, *Magnetic fields in massive stars. II. The buoyant rise of magnetic flux tubes through the radiative interior*, Astrophys. J., **586**, 480–494
MacDonald, J., & Mullan, D.J.: 2004, *Magnetic fields in massive stars: dynamics and origin*, Mon. Not. Roy. Astron. Soc., **348**, 702–716

On the inference of large-scale magnetic fields on slowly rotating chemically peculiar stars, see

Donati, J.-F., & Landstreet, J.D.: 2009, *Magnetic fields of nondegenerate stars*, Ann. Rev. Astron. Astrophys., **47**, 333–370

Figs. 5.5 and 5.6 were taken from, respectively

Mathys, G., Hubrig, S., Landstreet, J.D., Lanz, T., & Manfroid, J.: 1997, *The mean magnetic field modulus of AP stars*, Astron. & Astrophys. Suppl., **123**, 353–402
Silvester, J., Kochukhov, O., Wade, G.A., Piskunov, N., Landstreet, J.D., & Bagnulo, S.: 2009, *Cartography of the magnetic fields and chemical spots of Ap stars*, in *From Planets to Stars and Galaxies*, Strassmeier, K. G., Kosovichev, A. G., & Beckman, J. E., eds., IAU Symposium, 259, 403–404

On the stability of large-scale magnetic field in stellar radiative interiors, start with:

Pitts, E., & Tayler, R.J.: 1985, *The adiabatic stability of stars containing magnetic fields. IV-The influence of rotation*, Mon. Not. Roy. Astron. Soc., **216**, 139–154
Braithwaite, J., & Nordlund, Å.: 2006, *Stable magnetic fields in stellar interiors*, Astron. & Astrophys., **450**, 1077–1095

On the associated possible dynamo effect:

Spruit, H.C.: 2002, *Dynamo action by differential rotation in a stably stratified stellar interior*, Astron. & Astrophys, **381**, 923–932
Arlt, R., Hollerbach, R., & Rüdiger, G.: 2003, *Differential rotation decay in the radiative envelopes of CP stars*, Astron. & Astrophys., **401**, 1087–1094
Braithwaite, J.: 2006, *A differential rotation driven dynamo in a stably stratified star*, Astron. & Astrophys., **449**, 451–460

and for consequences on the structural, rotational and chemical evolution of massive stars:

Maeder, A., & Meynet, G.: 2003, *Stellar evolution with rotation and magnetic fields. I. The relative importance of rotational and magnetic effects*, Astron. & Astrophys., **411**, 543–552
Maeder, A., & Meynet, G.: 2005, *Stellar evolution with rotation and magnetic fields. III. The interplay of circulation and dynamo*, Astron. & Astrophys., **440**, 1041–1049

The potential observational consequences of iron opacity-driven outer convection zones in massive stars (including a brief discussion of the possibility of dynamo action) are discussed in

Cantiello, M., Langer, N., Brott, I., et al.: 2009, *Sub-surface convection zones in hot massive stars and their observable consequences*, Astron. & Astrophys., **499**, 279–290

5.7 Galaxies and Beyond

Observations of solar-like activity in late-type stars is the subject of the following two recent online review articles:

Berdyugina, S.V.: 2005, *Starspots: a key to the stellar dynamo*, Living Reviews in Solar Physics, **2**, 8, http://solarphysics.livingreviews.org/Articles/lrsp-2005-8/

Hall, J. C.: 2008, *Stellar chromospheric activity*, Living Reviews in Solar Physics, **5**, 2, http://solarphysics.livingreviews.org/Articles/lrsp-2008-2/

Figure 5.7 was taken from the following paper, still today one of the more cogent exposition of the Mt. Wilson CaII project and data:

Baliunas, S.L., Donahue, R.A., Soon, W.H., et al.: 1995, *Chromospheric variations in main-sequence stars*, Astrophys. J., **438**, 269–287

Figures 5.8 and 5.9 were taken from, respectively

Kraft, R.P.: 1967, *Studies of stellar rotation. V. The dependence of rotation on age among solar-type stars*, Astrophys. J., **150**, L183–L188

Skumanich, A.: 1972, *Time scales for Ca II emission decay, rotational braking, and Lithium depletion*, Astrophys. J., **171**, 565–567

On the confrontation of such observations with various types of dynamo models, start with

Noyes, R.W., Weiss, N.O., & Vaughan, A.H.: 1984, *The relation between stellar rotation rate and activity cycle periods*, Astrophys. J., **287**, 769–773

Saar, S.H., & Brandenburg, A.: 1999, *Time evolution of the magnetic activity cycle period. II. Results for an expanded stellar sample*, Astrophys. J., **524**, 295–310

Moss, D.: 2004, *Dynamo models and the flip-flop phenomenon in late-type stars*, Mon. Not. Roy. Astron. Soc., **352**, L17–L20

Jouve, L., Brown, B. P., & Brun, A. S.: 2010, *Exploring the P_{cyc} vs. P_{rot} relation with flux transport dynamo models of solar-like stars*, Astron. & Astrophys., **509**, A32

On the Weber-Davis MHD wind model, see

Weber, E.J., & Davis, Jr., L.: 1967, *The angular momentum of the solar wind*, Astrophys. J., **148**, 217–227

Belcher, J.W., & MacGregor, K.B.: 1976, *Magnetic acceleration of winds from solar-type stars*, Astrophys. J., **210**, 498–507

as well as

Keppens, R., & Goedbloed, J.P.: 1999, *Numerical simulations of stellar winds: polytropic models*, Astron. & Astrophys., **343**, 251–260

for a geometrically more realistic version. The idea that magnetized outflows can lead to stellar angular momentum loss can be traced to

Schatzman, E.: 1962, *A theory of the role of magnetic activity during star formation*, Annales d'Astrophysique, **25**, 18–29

In this context, another important pioneering paper is

Mestel, L.: 1968, *Magnetic braking by a stellar wind-I*, Mon. Not. Roy. Astron. Soc., **138**, 359–391

The theoretical derivation of Skumanich's square root relation in the context of $\alpha\Omega$ dynamo theory is due to

Durney, B.: 1972, *Evidence for changes in the angular velocity of the surface regions of the sun and stars-comments*, in *Solar Wind*, Sonett, C.P., Coleman, P.J., & Wilcox, J.M., eds., NASA Special Publication, 308, NASA, 282–286

There is a huge technical literature available on the observation, evolution and consequences of stellar rotation. The following conference proceedings volume should make a good starting point to those interested in digging further into the various ramifications of this topic:

Maeder, A., & Eenens, P., eds.: 2004, *Stellar rotation*, IAU Symposium, 215, Astronomical Society of the Pacific

The discussion of dynamo action in fully convective stars (Sect. 5.4) is based primarily on the following paper:

Dobler, W., Stix, M., & Brandenburg, A.: 2006, *Magnetic field generation in fully convective rotating spheres*, Astrophys. J., **638**, 336–347

but on this topic do not miss:

Durney, B.R., De Young, D.S., & Roxburgh, I.W.: 1993, *On the generation of the large-scale and turbulent magnetic fields in solar-type stars*, Solar Phys., **145**, 207–225
Browning, M.K.: 2008, *Simulations of dynamo action in fully convective stars*, Astrophys. J., **676**, 1262–1280

On magnetic field detection in white dwarfs and hot subdwarfs, see

Liebert, J., Bergeron, P., & Holberg, J.B.: 2003, *The true incidence of magnetism among field white dwarfs*, Astron. J., **125**, 348–353
Kawka, A., Vennes, S., Schmidt, G.D., Wickramasinghe, D.T., & Koch, R.: 2007, *Spectropolarimetric survey of Hydrogen-rich white dwarf stars*, Astrophys. J., **654**, 499–520
O'Toole, S.J., Jordan, S., Friedrich, S., & Heber, U.: 2005, *Discovery of magnetic fields in hot subdwarfs*, Astron. & Astrophys., **437**, 227–234

I found the following online article a good starting point to learn about galactic magnetic fields:

Beck, R.: 2007, *Galactic magnetic fields*, Scholarpedia, **2**, 2411, http://www.scholarpedia.org/article/Galactic_Magnetic_Fields

On galactic dynamo models, as well as alternatives explanatsions for galactic magnetic fields, the following review article provides an excellent entry point into the technical literature:

Kulsrud, R.M., & Zweibel, E.G.: 2008, *On the origin of cosmic magnetic fields*, Rep. Prog. Phys., **71**, 046901

Appendix A
Useful Identities and Theorems from Vector Calculus

A.1 Vector Identities

$$A \cdot (B \times C) = C \cdot (A \times B) = B \cdot (C \times A)$$
$$A \times (B \times C) = B(A \cdot C) - C(A \cdot B)$$
$$(A \times B) \times C = B(A \cdot C) - A(B \cdot C)$$
$$\nabla \times \nabla f = 0$$
$$\nabla \cdot (\nabla \times A) = 0$$
$$\nabla \cdot (fA) = (\nabla f) \cdot A + f(\nabla \cdot A)$$
$$\nabla \times (fA) = (\nabla f) \times A + f(\nabla \times A)$$
$$\nabla \cdot (A \times B) = B \cdot (\nabla \times A) - A \cdot (\nabla \times B)$$
$$\nabla(A \cdot B) = (B \cdot \nabla)A + (A \cdot \nabla)B + B \times (\nabla \times A) + A \times (\nabla \times B)$$
$$\nabla \cdot (AB) = (A \cdot \nabla)B + B(\nabla \cdot A)$$
$$\nabla \times (A \times B) = (B \cdot \nabla)A - (A \cdot \nabla)B - B(\nabla \cdot A) + A(\nabla \cdot B)$$
$$\nabla \times (\nabla \times A) = \nabla(\nabla \cdot A) - \nabla^2 A$$

A.2 The Gradient Theorem

For two points a, b in a space where a scalar function f with spatial derivatives everywhere well-defined up to first order,

$$\int_a^b (\nabla f) \cdot d\ell = f(b) - f(a),$$

independently of the integration path between a and b.

A.3 The Divergence Theorem

For any vector field A with spatial derivatives of all its scalar components everywhere well-defined up to first order,

$$\int_V (\nabla \cdot A) dV = \oint_S A \cdot \hat{n}\, dS ,$$

where the surface S encloses the volume V.

A.4 Stokes' Theorem

For any vector field A with spatial derivatives of all its scalar components everywhere well-defined up to first order,

$$\int_S (\nabla \times A) \cdot \hat{n}\, dS = \oint_\gamma A \cdot d\ell ,$$

where the contour γ delimits the surface S, and the orientation of the unit normal vector $\hat{n}$ and direction of contour integration are mutually linked by the right-hand rule.

A.5 Green's Identities

For any two scalar functions ϕ and ψ defined over a volume V bounded by a surface S and with spatial derivatives well-defined up to second order,

$$\int_V (\phi \nabla^2 \psi + \nabla \phi \cdot \nabla \psi) dV = \oint_S \phi(\nabla \psi) \cdot \hat{n}\, dS ,$$

and

$$\int_V (\phi \nabla^2 \psi - \psi \nabla^2 \phi) dV = \oint_S (\phi \nabla \psi - \psi \nabla \phi) \cdot \hat{n}\, dS .$$

These are known respectively as Green's first and second identities, the latter often simply referred to as Green's theorem.

Appendix B
Coordinate Systems and the Fluid Equations

This Appendix is adapted in part from Appendix B of the book by Jean-Louis Tassoul (1978) given in the bibliography to this appendix with a number of additions, including the MHD induction equation, expressions for the operators $\boldsymbol{u} \cdot \nabla$ and ∇^2 acting on a vector field, and for the divergence of a second rank tensor. Note also, in Sects. B.1.5 and B.2.5, the quantities in square brackets correspond to the components of the deformation tensor $D_{jk} = (1/2)(\partial_j u_k + \partial_k u_j)$.

B.1 Cylindrical Coordinates (s, ϕ, z)

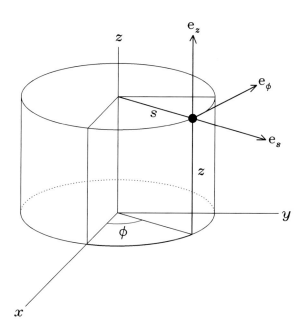

Fig. B.1 Geometric definition of cylindrical coordinates. The coordinate ranges are $s \in [0, \infty]$, $\phi \in [0, 2\pi]$, $z \in [-\infty, \infty]$. The cylindrical radius s is measured perpendicularly from the cartesian z-axis. The zero point of the azimuthal angle ϕ is on the cartesian x-axis. The local unit vector triad is oriented such that $\hat{\boldsymbol{e}}_z = \hat{\boldsymbol{e}}_s \times \hat{\boldsymbol{e}}_\phi$.

B.1.1 Conversion to Cartesian Coordinates

$$x = s\cos\phi, \quad y = s\sin\phi, \quad s = \sqrt{x^2 + y^2}, \quad \phi = \operatorname{atan}(y/x), \quad z = z.$$

$$\hat{e}_x = \cos\phi\,\hat{e}_s - \sin\phi\,\hat{e}_\phi, \qquad \hat{e}_y = \sin\phi\,\hat{e}_s + \cos\phi\,\hat{e}_\phi,$$

$$\hat{e}_s = \cos\phi\,\hat{e}_x + \sin\phi\,\hat{e}_y, \qquad \hat{e}_\phi = -\sin\phi\,\hat{e}_x + \cos\phi\,\hat{e}_y, \qquad \hat{e}_z = \hat{e}_z.$$

B.1.2 Infinitesimals

$$\mathrm{d}\boldsymbol{\ell} = \mathrm{d}s\,\hat{e}_s + s\,\mathrm{d}\phi\,\hat{e}_\phi + \mathrm{d}z\,\hat{e}_z$$

$$\mathrm{d}V = s\,\mathrm{d}s\,\mathrm{d}\phi\,\mathrm{d}z$$

B.1.3 Vector Operators

$$\frac{D}{Dt} = \frac{\partial}{\partial t} + u_s \frac{\partial}{\partial s} + \frac{u_\phi}{s}\frac{\partial}{\partial \phi} + u_z \frac{\partial}{\partial z}$$

$$\nabla f = \frac{\partial f}{\partial s}\hat{e}_s + \frac{1}{s}\frac{\partial f}{\partial \phi}\hat{e}_\phi + \frac{\partial f}{\partial z}\hat{e}_z$$

$$(\boldsymbol{u}\cdot\nabla)\boldsymbol{A} = \left(\boldsymbol{u}\cdot\nabla A_s - \frac{u_\phi A_\phi}{s}\right)\hat{e}_s + \left(\boldsymbol{u}\cdot\nabla A_\phi + \frac{u_\phi A_s}{s}\right)\hat{e}_\phi + (\boldsymbol{u}\cdot\nabla A_z)\hat{e}_z$$

$$\nabla\cdot\boldsymbol{A} = \frac{1}{s}\frac{\partial}{\partial s}(sA_s) + \frac{1}{s}\frac{\partial A_\phi}{\partial \phi} + \frac{\partial A_z}{\partial z}$$

$$\nabla\times\boldsymbol{A} = \left(\frac{1}{s}\frac{\partial A_z}{\partial \phi} - \frac{\partial A_\phi}{\partial z}\right)\hat{e}_s$$
$$+ \left(\frac{\partial A_s}{\partial z} - \frac{\partial A_z}{\partial s}\right)\hat{e}_\phi + \frac{1}{s}\left(\frac{\partial(sA_\phi)}{\partial s} - \frac{\partial A_s}{\partial \phi}\right)\hat{e}_z$$

$$\nabla^2 = \frac{1}{s}\frac{\partial}{\partial s}\left(s\frac{\partial}{\partial s}\right) + \frac{1}{s^2}\frac{\partial^2}{\partial \phi^2} + \frac{\partial^2}{\partial z^2}$$

$$\nabla^2 \boldsymbol{A} = \left(\nabla^2 A_s - \frac{A_s}{s^2} - \frac{2}{s^2}\frac{\partial A_\phi}{\partial \phi}\right)\hat{e}_s$$
$$+ \left(\nabla^2 A_\phi - \frac{A_\phi}{s^2} + \frac{2}{s^2}\frac{\partial A_s}{\partial \phi}\right)\hat{e}_\phi + \left(\nabla^2 A_z\right)\hat{e}_z$$

Appendix B: Coordinate Systems and the Fluid Equations   219

B.1.4 The Divergence of a Second-Order Tensor

$$[\nabla \cdot \mathbf{T}]_s = \frac{1}{s}\frac{\partial(sT_{ss})}{\partial s} + \frac{1}{s}\frac{\partial T_{\phi s}}{\partial \phi} + \frac{\partial T_{zs}}{\partial z} - \frac{T_{\phi\phi}}{s}$$

$$[\nabla \cdot \mathbf{T}]_\phi = \frac{1}{s}\frac{\partial(sT_{s\phi})}{\partial s} + \frac{1}{s}\frac{\partial T_{\phi\phi}}{\partial \phi} + \frac{\partial T_{z\phi}}{\partial z} + \frac{T_{\phi s}}{s}$$

$$[\nabla \cdot \mathbf{T}]_z = \frac{1}{s}\frac{\partial(sT_{sz})}{\partial s} + \frac{1}{s}\frac{\partial T_{\phi z}}{\partial \phi} + \frac{\partial T_{zz}}{\partial z}$$

B.1.5 Components of the Viscous Stress Tensor

$$\tau_{ss} = 2\mu\left[\frac{\partial u_s}{\partial s}\right] + \left(\zeta - \frac{2}{3}\mu\right)\nabla \cdot \mathbf{u}$$

$$\tau_{\phi\phi} = 2\mu\left[\frac{1}{s}\frac{\partial u_\phi}{\partial \phi} + \frac{u_s}{s}\right] + \left(\zeta - \frac{2}{3}\mu\right)\nabla \cdot \mathbf{u}$$

$$\tau_{zz} = 2\mu\left[\frac{\partial u_z}{\partial z}\right] + \left(\zeta - \frac{2}{3}\mu\right)\nabla \cdot \mathbf{u}$$

$$\tau_{s\phi} = \tau_{\phi s} = 2\mu\left[\frac{1}{2}\left(\frac{1}{s}\frac{\partial u_s}{\partial \phi} + s\frac{\partial}{\partial s}\frac{u_\phi}{s}\right)\right]$$

$$\tau_{\phi z} = \tau_{z\phi} = 2\mu\left[\frac{1}{2}\left(\frac{\partial u_\phi}{\partial z} + \frac{1}{s}\frac{\partial u_z}{\partial \phi}\right)\right]$$

$$\tau_{zs} = \tau_{sz} = 2\mu\left[\frac{1}{2}\left(\frac{\partial u_z}{\partial s} + \frac{\partial u_s}{\partial z}\right)\right]$$

B.1.6 Equations of Motion

$$\varrho\left(\frac{Du_s}{Dt} - \frac{u_\phi^2}{s}\right) = -\varrho\frac{\partial\Phi}{\partial s} - \frac{\partial p}{\partial s} + \frac{B_z}{\mu_0}\left(\frac{\partial B_s}{\partial z} - \frac{\partial B_z}{\partial s}\right)$$
$$- \frac{B_\phi}{\mu_0 s}\left(\frac{\partial(sB_\phi)}{\partial s} - \frac{\partial B_s}{\partial \phi}\right) + \frac{1}{s}\frac{\partial}{\partial s}(s\tau_{ss}) + \frac{1}{s}\frac{\partial \tau_{s\phi}}{\partial \phi} + \frac{\partial \tau_{sz}}{\partial z} - \frac{\tau_{\phi\phi}}{s}$$

$$\varrho\left(\frac{Du_\phi}{Dt} + \frac{u_\phi u_s}{s}\right) = -\frac{\varrho}{s}\frac{\partial\Phi}{\partial \phi} - \frac{1}{s}\frac{\partial p}{\partial \phi} + \frac{B_s}{\mu_0 s}\left(\frac{\partial(sB_\phi)}{\partial s} - \frac{\partial B_s}{\partial \phi}\right)$$
$$- \frac{B_z}{\mu_0}\left(\frac{1}{s}\frac{\partial B_z}{\partial \phi} - \frac{\partial B_\phi}{\partial z}\right) + \frac{1}{s}\frac{\partial}{\partial s}(s\tau_{\phi s}) + \frac{1}{s}\frac{\partial \tau_{\phi\phi}}{\partial \phi} + \frac{\partial \tau_{\phi z}}{\partial z} + \frac{\tau_{s\phi}}{s}$$

$$\varrho \frac{Du_z}{Dt} = -\varrho \frac{\partial \Phi}{\partial z} - \frac{\partial p}{\partial z} + \frac{B_\phi}{\mu_0}\left(\frac{1}{s}\frac{\partial B_z}{\partial \phi} - \frac{\partial B_\phi}{\partial z}\right)$$

$$- \frac{B_s}{\mu_0}\left(\frac{\partial B_s}{\partial z} - \frac{\partial B_z}{\partial s}\right) + \frac{1}{s}\frac{\partial}{\partial s}(s\tau_{zs}) + \frac{1}{s}\frac{\partial \tau_{z\phi}}{\partial \phi} + \frac{\partial \tau_{zz}}{\partial z}$$

B.1.7 The Energy Equation

$$\varrho T \frac{Ds}{Dt} = \Phi_\nu + \Phi_\eta + \frac{1}{s}\frac{\partial}{\partial s}\left[\chi s \frac{\partial T}{\partial s}\right] + \frac{1}{s^2}\frac{\partial}{\partial \phi}\left[\chi \frac{\partial T}{\partial \phi}\right] + \frac{\partial}{\partial z}\left[\chi \frac{\partial T}{\partial z}\right]$$

$$\Phi_\nu = 2\mu(D_{ss}^2 + D_{\phi\phi}^2 + D_{zz}^2 + 2D_{s\phi}^2 + 2D_{\phi z}^2 + 2D_{zs}^2) + (\zeta - \frac{2}{3}\mu)(\nabla \cdot \boldsymbol{u})^2$$

$$\Phi_\eta = \frac{\eta}{\mu_0}\left[\left(\frac{1}{s}\frac{\partial B_z}{\partial \phi} - \frac{\partial B_\phi}{\partial z}\right)^2 + \left(\frac{\partial B_s}{\partial z} - \frac{\partial B_z}{\partial s}\right)^2 + \frac{1}{s^2}\left(\frac{\partial (sB_\phi)}{\partial s} - \frac{\partial B_s}{\partial \phi}\right)^2\right]$$

B.1.8 The MHD Induction Equation

$$\frac{\partial B_s}{\partial t} = \frac{1}{s}\frac{\partial}{\partial \phi}(u_s B_\phi - u_\phi B_s) - \frac{\partial}{\partial z}(u_z B_s - u_s B_z)$$

$$- \frac{1}{s^2}\frac{\partial \eta}{\partial \phi}\left(\frac{\partial (sB_\phi)}{\partial s} - \frac{\partial B_s}{\partial \phi}\right) + \frac{\partial \eta}{\partial z}\left(\frac{\partial B_s}{\partial z} - \frac{\partial B_z}{\partial s}\right)$$

$$+ \eta\left(\nabla^2 B_s - \frac{B_s}{s^2} - \frac{2}{s^2}\frac{\partial B_\phi}{\partial \phi}\right)$$

$$\frac{\partial B_\phi}{\partial t} = \frac{\partial}{\partial z}(u_\phi B_z - u_z B_\phi) - \frac{\partial}{\partial s}(u_s B_\phi - u_\phi B_s)$$

$$- \frac{\partial \eta}{\partial z}\left(\frac{1}{s}\frac{\partial B_z}{\partial \phi} - \frac{\partial B_\phi}{\partial z}\right) + \frac{1}{s}\frac{\partial \eta}{\partial s}\left(\frac{\partial (sB_\phi)}{\partial s} - \frac{\partial B_s}{\partial \phi}\right)$$

$$+ \eta\left(\nabla^2 B_\phi - \frac{B_\phi}{s^2} + \frac{2}{s^2}\frac{\partial B_s}{\partial \phi}\right)$$

$$\frac{\partial B_z}{\partial t} = \frac{1}{s}\frac{\partial}{\partial s}(su_z B_s - su_s B_z) - \frac{1}{s}\frac{\partial}{\partial \phi}(u_\phi B_z - u_z B_\phi)$$

$$- \frac{\partial \eta}{\partial s}\left(\frac{\partial B_s}{\partial z} - \frac{\partial B_z}{\partial s}\right) + \frac{1}{s}\frac{\partial \eta}{\partial \phi}\left(\frac{1}{s}\frac{\partial B_z}{\partial \phi} - \frac{\partial B_\phi}{\partial z}\right) + \eta\left(\nabla^2 B_z\right)$$

B.2 Spherical Coordinates (r, θ, ϕ)

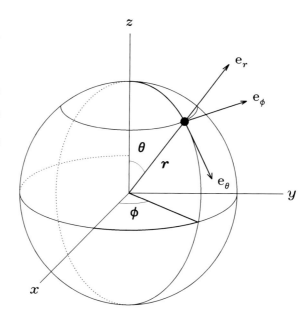

Fig. B.2 Geometric definition of polar spherical coordinates. The coordinate ranges are $r \in [0, \infty]$, $\theta \in [0, \pi]$, $\phi \in [0, 2\pi]$. The zero point of the azimuthal angle ϕ is on the cartesian x-axis and the zero point of the polar angle θ (sometimes called *colatitude*) is on the cartesian z-axis. Note that in so-called geographical coordinates, $longitude \equiv \phi$, but $latitude \equiv \pi/2 - \theta$. The local unit vector triad is oriented such that $\hat{e}_r = \hat{e}_\theta \times \hat{e}_\phi$.

B.2.1 Conversion to Cartesian Coordinates

$$x = r\sin\theta\cos\phi, \quad y = r\sin\theta\sin\phi, \quad z = r\cos\theta.$$
$$r = \sqrt{x^2 + y^2 + z^2}, \quad \theta = \mathrm{atan}\left(\sqrt{x^2+y^2}/z\right), \quad \phi = \mathrm{atan}(y/x).$$
$$\hat{e}_x = \sin\theta\cos\phi\,\hat{e}_r + \cos\theta\cos\phi\,\hat{e}_\theta - \sin\phi\,\hat{e}_\phi,$$
$$\hat{e}_y = \sin\theta\sin\phi\,\hat{e}_r + \cos\theta\sin\phi\,\hat{e}_\theta + \cos\phi\,\hat{e}_\phi,$$
$$\hat{e}_z = \cos\theta\,\hat{e}_r - \sin\theta\,\hat{e}_\theta.$$
$$\hat{e}_r = \sin\theta\cos\phi\,\hat{e}_x + \sin\theta\sin\phi\,\hat{e}_y + \cos\theta\,\hat{e}_z,$$
$$\hat{e}_\theta = \cos\theta\cos\phi\,\hat{e}_x + \cos\theta\sin\phi\,\hat{e}_y - \sin\theta\,\hat{e}_z,$$
$$\hat{e}_\phi = -\sin\phi\,\hat{e}_x + \cos\phi\,\hat{e}_y.$$

B.2.2 Infinitesimals

$$\mathrm{d}\boldsymbol{\ell} = \mathrm{d}r\,\hat{e}_r + r\mathrm{d}\theta\,\hat{e}_\theta + r\sin\theta\,\mathrm{d}\phi\,\hat{e}_\phi$$
$$\mathrm{d}V = r^2\sin\theta\,\mathrm{d}r\,\mathrm{d}\theta\,\mathrm{d}\phi$$

B.2.3 Vector Operators

$$\frac{D}{Dt} = \frac{\partial}{\partial t} + u_r \frac{\partial}{\partial r} + \frac{u_\theta}{r} \frac{\partial}{\partial \theta} + \frac{u_\phi}{r \sin\theta} \frac{\partial}{\partial \phi}$$

$$\nabla f = \frac{\partial f}{\partial r} \hat{e}_r + \frac{1}{r} \frac{\partial f}{\partial \theta} \hat{e}_\theta + \frac{1}{r \sin\theta} \frac{\partial f}{\partial \phi} \hat{e}_\phi$$

$$(\boldsymbol{u} \cdot \nabla)\boldsymbol{A} = \left(\boldsymbol{u} \cdot \nabla A_r - \frac{u_\theta A_\theta}{r} - \frac{u_\phi A_\phi}{r} \right) \hat{e}_r$$
$$+ \left(\boldsymbol{u} \cdot \nabla A_\theta - \frac{u_\phi A_\phi}{r} \cot\theta + \frac{u_\theta A_r}{r} \right) \hat{e}_\theta$$
$$+ \left(\boldsymbol{u} \cdot \nabla A_\phi + \frac{u_\phi A_r}{r} + \frac{u_\phi A_\theta}{r} \cot\theta \right) \hat{e}_\phi$$

$$\nabla \cdot \boldsymbol{A} = \frac{1}{r^2} \frac{\partial}{\partial r}(r^2 A_r) + \frac{1}{r \sin\theta} \frac{\partial (A_\theta \sin\theta)}{\partial \theta} + \frac{1}{r \sin\theta} \frac{\partial A_\phi}{\partial \phi}$$

$$\nabla \times \boldsymbol{A} = \frac{1}{r \sin\theta} \left(\frac{\partial (A_\phi \sin\theta)}{\partial \theta} - \frac{\partial A_\theta}{\partial \phi} \right) \hat{e}_r$$
$$+ \frac{1}{r \sin\theta} \left(\frac{\partial A_r}{\partial \phi} - \frac{\partial (A_\phi r \sin\theta)}{\partial r} \right) \hat{e}_\theta + \frac{1}{r} \left(\frac{\partial (r A_\theta)}{\partial r} - \frac{\partial A_r}{\partial \theta} \right) \hat{e}_\phi$$

$$\nabla^2 = \frac{1}{r^2} \frac{\partial}{\partial r}\left(r^2 \frac{\partial}{\partial r} \right) + \frac{1}{r^2 \sin\theta} \frac{\partial}{\partial \theta}\left(\sin\theta \frac{\partial}{\partial \theta} \right) + \frac{1}{r^2 \sin^2\theta} \frac{\partial^2}{\partial \phi^2}$$

$$\nabla^2 \boldsymbol{A} = \left(\nabla^2 A_r - \frac{2 A_r}{r^2} - \frac{2}{r^2 \sin\theta} \frac{\partial A_\theta \sin\theta}{\partial \theta} - \frac{2}{r^2 \sin\theta} \frac{\partial A_\phi}{\partial \phi} \right) \hat{e}_r$$
$$+ \left(\nabla^2 A_\theta + \frac{2}{r^2} \frac{\partial A_r}{\partial \theta} - \frac{A_\theta}{r^2 \sin^2\theta} - \frac{2 \cos\theta}{r^2 \sin^2\theta} \frac{\partial A_\phi}{\partial \phi} \right) \hat{e}_\theta$$
$$+ \left(\nabla^2 A_\phi + \frac{2}{r^2 \sin\theta} \frac{\partial A_r}{\partial \phi} + \frac{2 \cos\theta}{r^2 \sin^2\theta} \frac{\partial A_\theta}{\partial \phi} - \frac{A_\phi}{r^2 \sin^2\theta} \right) \hat{e}_\phi$$

B.2.4 The Divergence of a Second-Order Tensor

$$[\nabla \cdot \mathsf{T}]_r = \frac{1}{r^2} \frac{\partial (r^2 T_{rr})}{\partial r} + \frac{1}{r \sin\theta} \frac{\partial (T_{\theta r} \sin\theta)}{\partial \theta} + \frac{1}{r \sin\theta} \frac{\partial T_{\phi r}}{\partial \phi} - \frac{T_{\theta\theta} + T_{\phi\phi}}{r}$$

$$[\nabla \cdot \mathsf{T}]_\theta = \frac{1}{r^2} \frac{\partial (r^2 T_{r\theta})}{\partial r} + \frac{1}{r \sin\theta} \frac{\partial (T_{\theta\theta} \sin\theta)}{\partial \theta} + \frac{1}{r \sin\theta} \frac{\partial T_{\phi\theta}}{\partial \phi} + \frac{T_{\theta r}}{r} - \frac{T_{\phi\phi} \cot\theta}{r}$$

$$[\nabla \cdot \mathsf{T}]_\phi = \frac{1}{r^2} \frac{\partial (r^2 T_{r\phi})}{\partial r} + \frac{1}{r \sin\theta} \frac{\partial (T_{\theta\phi} \sin\theta)}{\partial \theta} + \frac{1}{r \sin\theta} \frac{\partial T_{\phi\phi}}{\partial \phi} + \frac{T_{\phi r}}{r} + \frac{T_{\theta\phi} \cot\theta}{r}$$

Appendix B: Coordinate Systems and the Fluid Equations

B.2.5 Components of the Viscous Stress Tensor

$$\tau_{rr} = 2\mu \left[\frac{\partial u_r}{\partial r}\right] + \left(\zeta - \frac{2}{3}\mu\right) \nabla \cdot \mathbf{u}$$

$$\tau_{\theta\theta} = 2\mu \left[\frac{1}{r}\frac{\partial u_\theta}{\partial \theta} + \frac{u_r}{r}\right] + \left(\zeta - \frac{2}{3}\mu\right) \nabla \cdot \mathbf{u}$$

$$\tau_{\phi\phi} = 2\mu \left[\frac{1}{r\sin\theta}\frac{\partial u_\phi}{\partial \phi} + \frac{u_r}{r} + \frac{u_\theta \cot\theta}{r}\right] + \left(\zeta - \frac{2}{3}\mu\right) \nabla \cdot \mathbf{u}$$

$$\tau_{r\theta} = \tau_{\theta r} = 2\mu \left[\frac{1}{2}\left(\frac{1}{r}\frac{\partial u_r}{\partial \theta} + r\frac{\partial}{\partial r}\frac{u_\theta}{r}\right)\right]$$

$$\tau_{\theta\phi} = \tau_{\phi\theta} = 2\mu \left[\frac{1}{2}\left(\frac{1}{r\sin\theta}\frac{\partial u_\theta}{\partial \phi} + \frac{\sin\theta}{r}\frac{\partial}{\partial \theta}\frac{u_\phi}{\sin\theta}\right)\right]$$

$$\tau_{\phi r} = \tau_{r\phi} = 2\mu \left[\frac{1}{2}\left(r\frac{\partial}{\partial r}\frac{u_\phi}{r} + \frac{1}{r\sin\theta}\frac{\partial u_r}{\partial \phi}\right)\right]$$

B.2.6 Equations of Motion

$$\varrho\left(\frac{Du_r}{Dt} - \frac{u_\theta^2 + u_\phi^2}{r}\right) = -\varrho\frac{\partial \Phi}{\partial r} - \frac{\partial p}{\partial r}$$
$$+ \frac{B_\phi}{\mu_0 r \sin\theta}\left(\frac{\partial B_r}{\partial \phi} - \frac{\partial}{\partial r}(B_\phi r \sin\theta)\right) - \frac{B_\theta}{\mu_0 r}\left(\frac{\partial (rB_\theta)}{\partial r} - \frac{\partial B_r}{\partial \theta}\right)$$
$$+ \frac{1}{r\sin\theta}\left[\frac{\sin\theta}{r}\frac{\partial}{\partial r}(r^2 \tau_{rr}) + \frac{\partial}{\partial \theta}(\tau_{r\theta}\sin\theta) + \frac{\partial \tau_{r\phi}}{\partial \phi}\right] - \frac{\tau_{\theta\theta} + \tau_{\phi\phi}}{r}$$

$$\varrho\left(\frac{Du_\theta}{Dt} + \frac{u_r u_\theta}{r} - \frac{u_\phi^2 \cot\theta}{r}\right) = -\frac{\varrho}{r}\frac{\partial \Phi}{\partial \theta} - \frac{1}{r}\frac{\partial p}{\partial \theta}$$
$$+ \frac{B_r}{\mu_0 r}\left(\frac{\partial (rB_\theta)}{\partial r} - \frac{\partial B_r}{\partial \theta}\right) - \frac{B_\phi}{\mu_0 r \sin\theta}\left(\frac{\partial (B_\phi \sin\theta)}{\partial \theta} - \frac{\partial B_\theta}{\partial \phi}\right)$$
$$+ \frac{1}{r\sin\theta}\left[\frac{\sin\theta}{r}\frac{\partial}{\partial r}(r^2 \tau_{\theta r}) + \frac{\partial}{\partial \theta}(\tau_{\theta\theta}\sin\theta) + \frac{\partial \tau_{\theta\phi}}{\partial \phi}\right] + \frac{\tau_{r\theta}}{r} - \frac{\tau_{\phi\phi}\cot\theta}{r}$$

$$\varrho\left(\frac{Du_\phi}{Dt} + \frac{u_r u_\phi}{r} + \frac{u_\theta u_\phi \cot\theta}{r}\right) = -\frac{\varrho}{r\sin\theta}\frac{\partial \Phi}{\partial \phi} - \frac{1}{r\sin\theta}\frac{\partial p}{\partial \phi}$$
$$+ \frac{B_\theta}{\mu_0 r \sin\theta}\left(\frac{\partial (B_\phi \sin\theta)}{\partial \theta} - \frac{\partial B_\theta}{\partial \phi}\right) - \frac{B_r}{\mu_0 r \sin\theta}\left(\frac{\partial B_r}{\partial \phi} - \frac{\partial (B_\phi r \sin\theta)}{\partial r}\right)$$
$$+ \frac{1}{r\sin\theta}\left[\frac{\sin\theta}{r}\frac{\partial}{\partial r}(r^2 \tau_{\phi r}) + \frac{\partial}{\partial \theta}(\tau_{\phi\theta}\sin\theta) + \frac{\partial \tau_{\phi\phi}}{\partial \phi}\right] + \frac{\tau_{r\phi}}{r} + \frac{\tau_{\theta\phi}\cot\theta}{r}$$

B.2.7 The Energy Equation

$$\varrho T \frac{Ds}{Dt} = \Phi_\nu + \Phi_\eta + \frac{1}{r^2}\frac{\partial}{\partial r}\left[\chi r^2 \frac{\partial T}{\partial r}\right]$$

$$+ \frac{1}{r^2 \sin\theta}\frac{\partial}{\partial \theta}\left[\chi \sin\theta \frac{\partial T}{\partial \theta}\right] + \frac{1}{r^2 \sin^2\theta}\frac{\partial}{\partial \phi}\left[\chi \frac{\partial T}{\partial \phi}\right]$$

$$\Phi_\nu = 2\mu(D_{rr}^2 + D_{\theta\theta}^2 + D_{\phi\phi}^2 + 2D_{r\theta}^2 + 2D_{\theta\phi}^2 + 2D_{\phi r}^2) + \left(\zeta - \frac{2}{3}\mu\right)(\nabla \cdot \boldsymbol{u})^2$$

$$\Phi_\eta = \frac{\eta}{\mu_0 r^2 \sin^2\theta}\left[\left(\frac{\partial(B_\phi \sin\theta)}{\partial \theta} - \frac{\partial B_\theta}{\partial \phi}\right)^2 \right.$$

$$\left. + \left(\frac{\partial B_r}{\partial \phi} - \frac{\partial(B_\phi r \sin\theta)}{\partial r}\right)^2 + \sin^2\theta\left(\frac{\partial(r B_\theta)}{\partial r} - \frac{\partial B_r}{\partial \theta}\right)^2\right]$$

B.2.8 The MHD Induction Equation

$$\frac{\partial B_r}{\partial t} = \frac{1}{r \sin\theta}\left[\frac{\partial}{\partial \theta}(\sin\theta (u_r B_\theta - u_\theta B_r)) - \frac{\partial}{\partial \phi}(u_\phi B_r - u_r B_\phi)\right]$$

$$- \frac{1}{r^2}\frac{\partial \eta}{\partial \theta}\left(\frac{\partial(r B_\theta)}{\partial r} - \frac{\partial B_r}{\partial \theta}\right) + \frac{1}{r^2 \sin^2\theta}\frac{\partial \eta}{\partial \phi}\left(\frac{\partial B_r}{\partial \phi} - \frac{\partial(B_\phi r \sin\theta)}{\partial r}\right)$$

$$+ \eta\left(\nabla^2 B_r - \frac{2 B_r}{r^2} - \frac{2}{r^2 \sin\theta}\frac{\partial(B_\theta \sin\theta)}{\partial \theta} - \frac{2}{r^2 \sin\theta}\frac{\partial B_\phi}{\partial \phi}\right)$$

$$\frac{\partial B_\theta}{\partial t} = \frac{1}{r \sin\theta}\frac{\partial}{\partial \phi}(u_\theta B_\phi - u_\phi B_\theta) - \frac{1}{r}\frac{\partial}{\partial r}(r u_r B_\theta - r u_\theta B_r)$$

$$- \frac{1}{r^2 \sin^2\theta}\frac{\partial \eta}{\partial \phi}\left(\frac{\partial(B_\phi \sin\theta)}{\partial \theta} - \frac{\partial B_\theta}{\partial \phi}\right) + \frac{1}{r}\frac{\partial \eta}{\partial r}\left(\frac{\partial(r B_\theta)}{\partial r} - \frac{\partial B_r}{\partial \theta}\right)$$

$$+ \eta\left(\nabla^2 B_\theta + \frac{2}{r^2}\frac{\partial B_r}{\partial \theta} - \frac{B_\theta}{r^2 \sin^2\theta} - \frac{2\cos\theta}{r^2 \sin^2\theta}\frac{\partial B_\phi}{\partial \phi}\right)$$

$$\frac{\partial B_\phi}{\partial t} = \frac{1}{r}\left[\frac{\partial}{\partial r}(r u_\phi B_r - r u_r B_\phi) - \frac{\partial}{\partial \theta}(u_\theta B_\phi - u_\phi B_\theta)\right]$$

$$- \frac{1}{r \sin\theta}\frac{\partial \eta}{\partial r}\left(\frac{\partial B_r}{\partial \phi} - \frac{\partial(B_\phi r \sin\theta)}{\partial r}\right) + \frac{1}{r^2 \sin\theta}\frac{\partial \eta}{\partial \theta}\left(\frac{\partial(B_\phi \sin\theta)}{\partial \theta} - \frac{\partial B_\theta}{\partial \phi}\right)$$

$$+ \eta\left(\nabla^2 B_\phi + \frac{2}{r^2 \sin\theta}\frac{\partial B_r}{\partial \phi} + \frac{2\cos\theta}{r^2 \sin^2\theta}\frac{\partial B_\theta}{\partial \phi} - \frac{B_\phi}{r^2 \sin^2\theta}\right)$$

Bibliography

For more on all this stuff, see

Morse, P. M., & Feshbach, H.: 1953, *Methods of theoretical physics*, McGraw-Hill, chap. 1

Batchelor, G. K.: 1967, *An introduction to fluid dynamics*. Cambridge University Press, appendix 2

Arfken, G. B.: 1970, *Mathematical methods for physicists*, 2nd edn., Academic Press, chap. 2

Tassoul, J.-L.: 1978, *Theory of rotating stars*. Princeton University Press, appendix B

or, for a concise introduction, Appendix A.2 of

Goedbloed, H., & Poedts, S.: 2004, *Principles of magnetohydrodynamics*, Cambridge University Press

Appendix C
Physical and Astronomical Constants

C.1 Physical Constants

Physical quantity	Symbol	Value	Units (SI)
Charge of electron	e	1.602×10^{-19}	C
Mass of electron	m_e	9.109×10^{-31}	kg
Mass of proton	m_p	1.673×10^{-27}	kg
Permittivity of vacuum	ε_0	8.854×10^{-12}	$C^2 N^{-1} m^{-2}$
Permeability of vacuum	μ_0	$4\pi \times 10^{-7}$	$N A^{-2}$
Speed of light	c	2.998×10^{8}	$m\,s^{-1}$
Planck constant	h	6.626×10^{-34}	$J\,s$
Boltzmann constant	k	1.381×10^{-23}	$J\,K^{-1}$
Stefan–Boltzmann constant	σ	5.670×10^{-8}	$J\,K^{-4} m^{-2} s^{-1}$
Gravitational constant	G	6.673×10^{-11}	$m^3 kg^{-1} s^{-2}$

C.2 Astronomical Constants

Astronomical quantity	Symbol	Value	Units (SI)
Earth mass	$M_\oplus$	5.972×10^{24}	kg
Earth radius	$R_\oplus$	6.378×10^{6}	m
Astronomical unit	AU	1.496×10^{11}	m
Solar mass	$M_\odot$	1.989×10^{30}	kg
Solar radius	$R_\odot$	6.960×10^{8}	m
Solar luminosity	$L_\odot$	3.84×10^{26}	$J\,s^{-1}$
Parsec	pc	3.086×10^{16}	m
Light-year	ly	9.461×10^{15}	m

Appendix D
Maxwell's Equations and Physical Units

Electromagnetism is, unfortunately, a subfield of physics where the choice of units does not only influence the numerical values assigned to measurements, but also the mathematical form of the fundamental laws, i.e., Maxwell's equations.

D.1 Maxwell's Equations

The whole mess in converting SI units to the astrophysically ubiquitous CGS units all harks back to the definition for the unit of charge, as embodied in Coulomb's law. Under the SI system we write the electrostatic force between two charges q_1 and q_2 located at positions x_1 and x_2 as

$$F = \frac{1}{4\pi\varepsilon_0} \frac{q_1 q_2}{r^2} \hat{r} \quad [\text{SI}], \tag{D.1}$$

with electrical charge measured in coulomb, and with $r \equiv x_1 - x_2$ for notational brevity; whereas under the CGS system the constant $1/4\pi\varepsilon_0$ is absorbed into the definition of the unit of charge:

$$F = \frac{q_1 q_2}{r^2} \hat{r} \quad [\text{CGS}], \tag{D.2}$$

with electrical charge now measured in "electrostatic units", abbreviated "esu" and sometimes also called "statcoulomb". It electrostatics it is relatively easy to switch from CGS to SI with the simple substitution $\varepsilon_0 \to 1/(4\pi)$. With electrical currents now measured in esu s^{-1} in the CGS system, and remembering that $c^2 = (\varepsilon_0 \mu_0)^{-1}$, the $\mu_0/4\pi$ prefactor in the Biot–Savart law now becomes $1/c$:

$$B = \frac{1}{c} \int \frac{I d\boldsymbol{\ell} \times \hat{r}}{r^2} \quad [\text{CGS}] \quad [\text{Biot–Savart}]. \tag{D.3}$$

If you then now go through the process of re-constructing Maxwell's equations under these two new forms for the fundamental relations (electric and magnetic forces), you eventually get to

$$\nabla \cdot \boldsymbol{E} = 4\pi \varrho_e \quad \text{[Gauss' law]}, \tag{D.4}$$

$$\nabla \cdot \boldsymbol{B} = 0 \quad \text{[Anonymous]}, \tag{D.5}$$

$$\nabla \times \boldsymbol{E} = -\frac{1}{c}\frac{\partial \boldsymbol{B}}{\partial t} \quad \text{[Faraday's law]}, \tag{D.6}$$

$$\nabla \times \boldsymbol{B} = \frac{4\pi}{c}\boldsymbol{J} + \frac{1}{c}\frac{\partial \boldsymbol{E}}{\partial t} \quad \text{[Ampere's /Maxwell's law]}. \tag{D.7}$$

In some sense, the CGS system is perhaps more "natural", as it omits the introduction of new, apparently fundamental physical constants ε_0 and μ_0, to simply stick with the speed of light c, the only price to pay being an extraneous factor 4π in Gauss' law. The Lorentz force and Poynting vector become, in CGS units:

$$\boldsymbol{F} = q\left(\boldsymbol{E} + \frac{1}{c}\boldsymbol{u} \times \boldsymbol{B}\right) \quad \text{[Lorentz force]}, \tag{D.8}$$

$$\boldsymbol{S} = \frac{c}{4\pi}(\boldsymbol{E} \times \boldsymbol{B}) \quad \text{[Poynting flux]}, \tag{D.9}$$

and the electrostatic and magnetic energies:

$$\mathcal{E}_e = \frac{1}{8\pi}\int \mathrm{E}^2 dV, \tag{D.10}$$

$$\mathcal{E}_B = \frac{1}{8\pi}\int \mathrm{B}^2 dV. \tag{D.11}$$

D.2 Conversion of Units

Table D.1 gives you the conversion factor (f) required to go *from* SI *to* cgs units, i.e., the numerical value of a quantity in cgs Unit $= f \times$ numerical value in SI units. Any "3" appearing in a given value for f is a notational shortcut for 2.99792458.

For a somewhat humourous close to this rather dry Appendix, here are five different ways, actually to be found in various textbooks or research monographs, to express teslas in terms of other fundamental SI units:

$$1\,\mathrm{T} = 1\,\frac{\mathrm{V\,s}}{\mathrm{m}^2} = 1\,\frac{\mathrm{N}}{\mathrm{A\,m}} = 1\,\frac{\mathrm{kg}}{\mathrm{A\,s}^2} = 1\,\frac{\mathrm{Wb}}{\mathrm{m}^2} = 1\,\frac{\mathrm{kg}}{\mathrm{C\,s}}. \tag{D.12}$$

Appendix D: Maxwell's Equations and Physical Units

Table D.1 Conversion between SI ands CGS units

Quantity	SI name	SI symbol	Conv. factor f	CGS name	CGS symbol
Length	meter	m	10^2	centimeter	cm
Mass	kilogram	kg	10^3	gram	g
Force	newton	N	10^5	dyne	dyne
Energy	joule	J	10^7	erg	erg
Charge	coulomb	C	3×10^9	electrostatic units	esu
Current	ampere	A	3×10^9	statampere	esu s^{-1}
Potential	volt	V	1/300	statvolt	statvolt
Electric field	—	V m^{-1}	$(1/3) \times 10^{-4}$	—	statvolt cm^{-1}
Magnetic field	tesla	T	10^4	gauss	G
Magnetic flux	weber	Wb	10^8	maxwell	Mx

Bibliography

The bulk of this appendix is taken from Appendix C in

Griffiths, D. J.: 1999, *Introduction to electrodynamics*, 3rd edn., Prentice-Hall

with some additions extracted primarily from:

Huba, J. D.: 2011, *NRL Plasma Formulary* No. NRL/PU/6790-11-551, Naval Research Laboratory

Index

Λ-quenching. *See* Differential rotation:
  Λ-quenching
Ω-effect, 104. *See also* Differential rotation
Ω-loop, 90, 121
α-effect, 97–101, 103, 104
  and current helicity, 101
  and kinetic helicity, 100, 141
  catastrophic quenching, 119
  cyclonic fluid motions, 98
  flux tube, 134, 175
  hemispheric dependency, 100, 108, 114, 134
  MHD simulations, 101, 141
  parametrization, 108, 114, 115, 119, 189
  quenching, 101, 112, 113, 119, 138, 168
  in tachocline, 133
α-tensor, 97, 100, 101, 141

A
Alfvén radius, 204
Alfvén's theorem, 25
Alfvén velocity, 22, 204
Alfvén wave, 22
Aly's theorem, 27
Ampére's law, 13, 14
Anelastic approximation, 22, 136, 143
Anti-dynamo theorems, 64–68
Axisymmetric magnetic field, 38, 66, 141
Axisymmetrization, 62–64, 127

B
Babcock Harold D., 89
Babcock Horace W., 89, 121
Babcock–Leighton mechanism, 121–127, 188
Battery mechanism, 29
Beltrami flow, 68

Bifurcation diagram, 172
BMR. *See* Solar magnetic field: bipolar magnetic region
Boussinesq approximation, 21
Butterfly diagram. *See* Sunspots: butterfly diagram

C
Calcium index, 198, 202
Carrington Richard, 156
Chaotic trajectories, 78, 82
Charge relaxation time, 18, 28
Chinese imperial courts, 154
Compton drag effect, 30
Conductivity
  electrical, 14
  radiative, 12
  thermal, 12
Constructive folding, 48, 80, 103, 144
Continuity equation, 6
Coriolis force, 94, 108, 122
Coriolis number, 100, 200
Cosmogenic radioisotopes, 161, 173
Coulomb gauge, 26
Cowling's theorem, 66, 68, 106, 130
CP (circularly polarized) flow, 75–77, 79, 82
Critical dynamo number, 33, 110, 172, 173
Current helicity, 26, 101, 141, 142

D
Dalton minimum, 161, 165
Data assimilation, 181
Deformation tensor, 9
Democritus, 2

D (*cont.*)
Destructive folding, 48, 55, 59, 75, 103, 144
Differential rotation
  Λ-quenching, 168–170
  and dynamo waves, 105, 111, 112, 115
  early-type stars, 192, 194
  from numerical simulation, 140, 192
  inductive action, 48, 49, 51, 118, 132
  magnetic backreaction, 138, 140, 168
  solar, 51, 140
  solar parametrization, 50
  tachocline, 51, 93, 101, 118, 132, 140
  time-scale, 62, 64
Dipolar magnetic field, 41, 43, 92, 121, 122, 139, 191
  tilted, 62, 195
Dirac Paul A. M., 29
Dynamo
  $\alpha\Omega$, 107–121, 132, 163, 165–167, 179, 180, 206
  α^2, 106, 189, 190, 208
  $\alpha^2\Omega$, 107, 143, 191, 192
  amplitude modulation, 115, 165, 169, 170, 173, 176
  Babcock–Leighton, 128–132, 165, 167, 172, 178–180, 206
  CP flow, 76–78, 81
  cycle period, 110, 113, 116, 118, 120, 130, 132, 200, 206
  energetics, 24, 33, 45, 52
  fast, 74–83, 143, 144
  fast versus slow, 74, 95, 143, 144
  flux transport, 116, 128, 133, 165, 170, 179
  growth rate, 33, 47, 71, 74, 76, 77, 109, 191
  homopolar, 32
  interface, 119
  interface $\alpha\Omega$, 119, 120
  kinematic. *See* Kinematic regime
  MHD simulation, 82, 136–145, 192, 193, 207
  non-kinematic, 168
  Roberts cell, 70, 73
  saturation, 51, 75, 112, 118, 138
  stochastic forcing, 164, 167, 176
  stretch–twist–fold, 47, 52, 75
  time delay, 128, 132, 171, 176
  turbulent, 82, 136
Dynamo equations, 103, 104, 106
  α^2 versus $\alpha\Omega$ versus $\alpha^2\Omega$, 106
  axisymmetric, 106
  boundary conditions, 107, 114
Dynamo number, 106, 165, 171. *See also* Critical dynamo number

Dynamo problem, 33, 34
Dynamo waves, 105, 112, 114, 116, 119, 192.
  See also Differential rotation: and dynamo waves

E
Eigenmode
  $\alpha\Omega$, 109–111
  α^2, 189, 190
  $\alpha^2\Omega$, 191, 192
  diffusive, 40
  Roberts cell, 70
Electric field, 13, 14, 27
  astrophysical, 27, 28
Electrical charge conservation, 18, 19
Electrical current density, 17
Electromagnetic skin depth, 59, 111, 120, 192
Electromotive force, 24, 29, 31. *See also* Mean electromotive force
emf. *See* Electromotive force
Entropy equation, 12
Equipartition field strength, 112, 118, 192, 207
Exponential stretching, 72

F
Fabricius Johann, 154
Faraday Michael, 30
Faraday's law, 13
Fast dynamo. *See* Dynamo: fast
Fast mode, 22
Ferraro's theorem, 50
First order smoothing approximation, 99
Fluctuation
  α-effect, 165
  coherence time, 165
  in meridional flow, 170
Fluid
  as continuum, 2, 18
  circulation, 11
  density, 2
  entropy, 12
  incompressible, 6, 9
  internal energy, 12
  mass conservation, 6, 45
  Newtonian, 9
  pressure, 8
  stress, 3
  thermal dilation, 6
  vorticity, 11, 68
Flux expulsion, 58, 70, 82, 144
Flux freezing, 25, 45, 55, 56, 210

Index 235

Force-free magnetic field, 26, 27
FOSA. *See* First order smoothing approximation
Fossil magnetic field. *See* Stellar magnetic fields: fossil

G
Galileo Galilei, 154
Gauss' law, 13
Gleissberg cycle, 158, 169
Gnevyshev–Ohl rule, 158, 165, 166
Goldsmid Johann a.k.a. Fabricius, 154
Granulation, 144

H
Hale George E., 88, 90, 91
Hale's polarity rules, 90, 121
Harriot Thomas, 154
Helicity. *See* Kinetic helicity, Current helicity, Magnetic helicity
Hydrodynamical shearing, 48–50, 55
Hydrodynamical stretching, 44, 47, 72

I
Induction equation. *See* MHD: induction equation
Intermittency, 138, 173–176, 200
 spatial, 76, 82
 temporal, 77
Iterative map, 171

J
Joule heating, 20
Joy's law, 91, 94, 121

K
Kelvin's theorem, 11
Kepler Johannes, 154
Kinematic regime, 33, 49, 107, 113, 167, 173, 189, 207
Kinematic theorem, 25
Kinetic helicity, 12, 68, 100, 141, 142, 207

L
Lagrangian derivative, 7
Lorentz force, 18, 19, 22, 31, 45, 51, 112, 138, 168
Lyapunov exponent, 78

M
Magnetic boundary conditions, 60
Magnetic boundary layers, 59, 70
Magnetic diffusion time, 16, 28, 38, 41, 42, 61, 102, 108, 123, 194
 local versus global, 57
Magnetic diffusivity, 15
 parametrization, 42, 189
Magnetic energy, 24, 27, 54, 113, 136, 207
Magnetic flux ropes, 90, 93, 134, 162, 175, 194
Magnetic helicity, 26, 74, 75
Magnetic monopoles, 29
Magnetic network, 89
Magnetic Prandtl number, 17, 169
Magnetic pressure, 22
Magnetic Reynolds number, 15–17, 58, 78, 106
Magnetic tension, 22
Magnetic vector potential, 25, 38, 66
Magnetic waves, 22
Magnetogram, 89
Magnetohydrodynamics. *See* MHD
Magnetosonic wave, 22
Malkus–Proctor effect, 140, 168
Material surface, 7
Maunder Annie, 157
Maunder E. Walter, 157
Maunder minimum, 160, 173, 200
Maxwell's equations, 13
Mean electromotive force, 96–98, 103, 141, 142
Mean-field dynamo models, 95
Mean-field electrodynamics, 95
Medieval maximum, 161
Meridional circulation, 116–118, 128, 171, 194
MHD, 1–34
 approximation, 13, 17, 19
 electrical charge density, 19
 governing equations, 20
 ideal limit, 16, 26, 58
 induction equation, 14, 25, 38, 49, 96
 instabilities, 132, 175
 Lorentz force. *See* Lorentz force
 singular limit, 16, 58

N
Navier–Stokes equation, 8
Newton Isaac, 7
Numerical simulations. *See* α-effect: MHD simulations, Differential rotation: from numerical simulation, Dynamo: MHD simulation

O

Ohm's law, 14
Ohmic dissipation, 15, 20, 26, 37, 58, 61

P

Parker Eugen N., 98, 119
Parker–Yoshimura sign rule, 112, 114, 116, 192
PDF. *See* Probability density function
Planar flow, 48, 52, 65
Poincaré section, 79
Poloidal/Toroidal separation, 38, 49, 66, 104, 106
Potential magnetic field, 27, 39
Poynting flux, 23, 60
Primordial relic, 28
Probability density function, 76, 78

Q

Quadrupolar magnetic field, 41, 208
Quenching. *See* α-effect: quenching, Differential rotation: Λ-quenching

R

Reynolds number, 10
  magnetic. *See* Magnetic Reynolds number
Roberts cell, 68, 73, 75, 82, 99
Roberts cell dynamo. *See* Dynamo: Roberts cell
Rossby number, 200

S

Scaling analysis, 15
Scheiner Christoph, 154
Schwabe Samuel H., 155
Second-order correlation approximation, 99, 100, 103, 141
Seed magnetic field, 28, 30, 33, 136, 211
Separation of scales, 2
Shearing. *See* Hydrodynamical shearing
Slow mode, 22
SOCA. *See* Second-order correlation approximation
Solar activity, 160
  proxy, 161, 198
Solar cycle, 89, 92, 153, 156, 157
  chaos, 173, 181
  forecasting, 176–182
  magnetic flux budget, 93
  polar fields, 92, 118, 121, 122, 179
  precursor, 177
Solar magnetic field
  bipolar magnetic region, 91, 121, 125
  large-scale, 90
  polarity reversal, 90, 92, 121, 122, 138
  small-scale, 89, 90, 95, 144
  surface evolution, 92, 93, 121, 124, 125, 139
Space climate, 176
Space weather, 153, 176
Spin-down. *See* Stars: spin-down
Spörer Gustav, 156
Spörer minimum, 161
SSN. *See* Sunspot: number
Stagnation point, 46, 68, 72
Stars *See also* Stellar magnetic fields
  convection, 187, 188, 194, 197
  early-type, 43
  fully convective, 188, 207
  rotation, 187, 200–202, 205
  solar type, 198
  spin–down, 195, 202, 205, 206, 209
  winds, 202
Stellar magnetic fields
  βCep sub-class, 195
  A and B stars, 195
  Ap/Bp stars, 43, 195, 196
  Doppler imaging, 195, 196, 208
  early A star, 193
  early-type, 189, 194, 195
  flux conservation, 43, 197
  fossil, 16, 43, 188, 197, 210
  late-type, 188, 198, 202, 207
  M-type stars, 207
  neutron stars, 43, 210
  stability, 197
  starspots, 198
  trapping in interior, 191, 193, 194
  white dwarfs, 43, 209
STF. *See* Dynamo: stretch–twist–fold
Stream function, 53, 68
Stress
  fluid, 3
  tensorial representation, 7
  viscous, 8
Stretching. *See* Hydrodynamical stretching
Sunspots
  as magnetic flux rope, 90
  butterfly diagram, 92, 115, 133, 135, 138, 156, 170, 174, 175
  cycle, 90, 155. *See also* Solar cycle
  cycle numbering convention, 155
  decay, 93, 122, 124, 125
  early telescopic observations, 154, 155
  leading, 90, 121, 124

magnetic nature, 88
naked-eye observations, 153
number, 155
pores, 144
proxy, 163–164
trailing, 90, 121, 124
Synoptic diagram, 92

T
Tachocline. *See* Differential rotation: tachocline
Taylor–Proudman theorem, 140
Theophrastus, 154
Thermal convection, 6, 82, 136, 144, 192, 207
Turbulent diffusivity, 102, 118, 119, 189
Turbulent pumping, 98, 101, 132, 138, 143
Turnover time, 16, 53, 58, 100, 102, 116

V
Viscosity
 bulk, 9
 dynamical, 9
 kinematic, 9
Viscous dissipation function, 13
Vorticity equation, 12
Vorticity tensor, 9

W
Waldmeier rule, 158, 166, 167
WD. *See* Weber–Davis
Weber–Davis MHD wind, 202, 204
Wolf minimum, 161
Wolf Rudolf, 155
Wolf sunspot number, 155
Worcester Chronicles, 154

Z
Zeeman broadening, 89
Zeeman splitting, 88, 195
Zeldovich's theorem, 65, 68

Made in the USA
Monee, IL
19 November 2023

46921822R00155